Kant and Kierkegaard on Time and Eternity

MERCER
UNIVERSITY PRESS

Endowed by
TOM WATSON BROWN
and
THE WATSON-BROWN FOUNDATION, INC.

Kant and Kierkegaard

on Time and Eternity

Ronald M. Green

MERCER UNIVERSITY PRESS
MACON, GEORGIA

MUP/H830

1400 Coleman Avenue
Macon, Georgia 31207

First Edition

Books published by Mercer University Press are printed on acid-free paper that meets the requirements of American National Standard for Information Sciences—Permanence of Paper for Printed Library Materials.

Mercer University Press is a member of Green Press Initiative (greenpressinitiative.org), a nonprofit organization working to help publishers and printers increase their use of recycled paper and decrease their use of fiber derived from endangered forests. This book is printed on recycled paper.
Library of Congress Cataloging-in-Publication Data

Green, Ronald Michael.
Kant and Kierkegaard on time and eternity / by Ronald M. Green. -- 1st ed.
p. cm.
Includes bibliographical references and index.
ISBN 978-0-88146-255-5 (hardcover : alk. paper)
1. Kierkegaard, S?ren, 1813-1855. 2. Kant, Immanuel, 1724-1804. 3. Time--Philosophy. 4. Eternity. I. Title.
B4377.G714 2011
198'.9--dc23

2011024532

Contents

Abbreviations

All references to Kant or Kierkegaard refer to the translations listed below. For Kant's *Foundations of the Metaphysics of Morals*, references are also included in brackets to pagination in the Königlich-Preussischen Akademie der Wissenschaften edition of the *Grundlegung zur Metaphysik der Sitten* (GR), vol. 4 in Kant's *Gesammelte Schriften.*

Kant

CF	*The Conflict of the Faculties/Der Streit der Facultäten*, trans. Mary J. Gregor (New York: Abaris Books, 1979)
D	*Dreams of a Spirit-Seer, and Other Related Writings*, trans. and commentary by John Manolesco (New York: Vantage Press, 1969)
GR	*Foundations of the Metaphysics of Morals*, trans. Lewis White Beck (Indianapolis, Indiana: Bobbs-Merrill, 1959)
KPV	*Critique of Practical Reason* [*Kritik der praktischen Vernunft*], trans. Lewis White Beck (Indianapolis, Indiana: Bobbs-Merrill, 1956)
KRV	*Critique of Pure Reason* [*Kritik der reinen Vernunft*], trans. Norman Kemp Smith (New York: St. Martin's Press, 1929)
KU	*Critique of Judgement* [*Kritik der Urteilskraft*], trans. J.H. Bernard (New York: Hafner Press, 1951)
LE	*Lectures on Ethics*, trans. Louis Infield (New York: Harper & Row, 1963)
LPT	*Lectures on Philosophical Theology*, trans. Allen Wood (Ithaca, New York: Cornell University Press, 1978)
MS	*Metaphysics of Morals* [*Metaphysik der Sitten*], trans.

	Mary Gregor (Cambridge: Cambridge University Press, 1991). This edition contains both the Doctrine of Right and the Doctrine of Virtue.
Rel	*Religion within the Limits of Reason Alone*, trans. Theodore M. Green and Hoyt H. Hudson (New York: Harper & Row, 1960)

Kierkegaard

CA	*The Concept of Anxiety*, trans. Reidar Thomte in collaboration with Albert B. Anderson (Princeton, New Jersey: Princeton University Press, 1980)
CUP	*Concluding Unscientific Postscript*, trans. Howard V. Hong and Edna H. Hong, 2 vols. (Princeton, New Jersey: Princeton University Press, 1992)
EO	*Either/Or* (1 and 2), trans. Howard V. Hong and Edna H. Hong, two vols. (Princeton, New Jersey: Princeton University Press, 1987)
FT	*Fear and Trembling* and *Repetition*, trans. Howard V. Hong and Edna H. Hong (Princeton, New Jersey: Princeton University Press, 1983)
JC	*Johannes Climacus or De omnibus dubitandum est.* See PF
JP	*Søren Kierkegaard's Journals and Papers*, ed. and trans. Howard V. Hong and Edna H. Hong, assisted by Gregor Malantschuk (Bloomington and London: Indiana University Press, 1967); 2nd ed. (1970); 3rd–4th ed. (1975); 5th–7th ed. (1978)
KAUC	*Kierkegaard's Attack upon "Christendom,"* 1854–1855, trans. Walter Lowrie (Princeton, New Jersey: Princeton University Press, 1944)
OAR	*On Authority and Revelation, The Book on Adler*, trans. Walter Lowrie (Princeton, New Jersey: Princeton University Press, 1955)
PAP	*Søren Kierkegaards Papirer*, ed. P.A. Heiberg and V. Kuhr, 16 vols. (Copenhagen, Denmark:

	Gyldendalske Boghandel Nordisk Verlag, 1909–1978)
PC	*Practice in Christianity*, trans. Howard V. Hong and Edna H. Hong (Princeton, New Jersey: Princeton University Press, 1991)
PF	*Philosophical Fragments* and "*Johannes Climacus eller De omnibus dubitandum est,*" trans. Howard V. Hong and Edna H. Hong (Princeton, New Jersey: Princeton University Press, 1985)
POV	*The Point of View on My Work as an Author*, ed. and trans. Howard V. Hong and Edna H. Hong (Princeton: Princeton University Press, 1998)
Rep	*Repetition*. See FT.
SLW	*Stages on Life's Way*, trans. Howard V. Hong and Edna H. Hong (Princeton, New Jersey: Princeton University Press, 1988)
SUD	*The Sickness unto Death*, trans. Howard V. Hong and Edna Hong (Princeton, New Jersey: Princeton University Press, 1980)
TA	*Two Ages: The Age of Revolution and the Present Age. A Literary Review*, trans. Howard V. Hong and Edna H. Hong (Princeton, New Jersey: Princeton University Press, 1978)
WL	*Works of Love*, trans. Howard V. Hong and Edna H. Hong (Princeton, New Jersey: Princeton University Press, 1995)

Chapter Sources

Chapters one through ten in this book are drawn from my previously published articles or chapters. Apart from typographical corrections, adjustments to Mercer University Press style, and changes made to avoid repetitions, the chapters here appear as they did in their original publication. Sources are listed in the order they appear in this book.

"The First Formulation of the Categorical Imperative as Literally a 'Legislative' Metaphor," *History of Philosophy Quarterly* 8, no. 2 (1991): 163–179. Reprinted with permission of *History of Philosophy Quarterly*.

"Kant on Christian Love," in *The Love Commandments: Essays in Christian Ethics and Moral Philosophy*, ed. William Werpehowski and Edmund Santurri (Washington, DC: Georgetown University Press, 1992) 261–280. Reprinted with permission of Georgetown University Press.

"The Limits of the Ethical in Kierkegaard's *The Concept of Anxiety* and Kant's *Religion within the Limits of Reason Alone*," in *International Kierkegaard Commentary, The Concept of Anxiety*, ed. Robert L. Perkins (Macon, Georgia: Mercer University Press, 1985) 63–87. Reprinted with permission of Mercer University Press.

"Kant and Kierkegaard on the Need for a Historical Faith: An Imaginary Dialogue," in *Kant and Kierkegaard on Religion*, ed. D.Z. Phillips and Timothy Tessin, Claremont Studies (Houndsmill, Basingstoke, Hampshire and London, England and New York: Macmillan and St. Martin's Press, 2000) 131–152. A revised version of this, on which chapter four is based, is reprinted in *Kant and the New Philosophy of Religion*, ed.

Christopher L. Firestone and Steven R. Palmquist (Bloomington: Indiana University Press, 2006) 157–175. Reprinted with permission of St. Martin's Press and Indiana University Press.

"Kant: A Debt both Obscure and Enormous," in *Kierkegaard and His German Contemporaries*, ed. Jon Stewart, Tome 1: Philosophy, in the series "Kierkegaard Research: Sources, Reception and Resources" (London: Ashgate, 2007) 179–210. Reprinted with permission.

"'Developing' *Fear and Trembling*," in *The Cambridge Companion to Kierkegaard*, ed. Alastair Hannay and Gordon Marino (Cambridge: Cambridge University Press, 1998) 257–281. Reprinted with permission of Cambridge University Press.

"*Fear and Trembling*: A Jewish Appreciation," in *Kierkegaard Studies*, ed. Niels Jørgen Cappelørn, Hermann Deuser, Jon Stewart, and Christian Fink Toistrup (Berlin / New York: Walter de Gruyter, 2002) 137–149. Reprinted with permission of Verlag Walter de Gruyter.

"Kierkegaard's Great Critique: *Either/Or* as a Kantian Transcendental Deduction," in *International Kierkegaard Commentary, Either/Or II*, ed. Robert L. Perkins (Macon, Georgia: Mercer University Press, 1995) 139–153. Reprinted with permission of Mercer University Press.

"*Either/Or*: Kierkegaard's Great Overture," in *Kierkegaard Studies Yearbook 2008*, ed. Niels Jørgen Cappelørn, Hermann Deuser, and K. Brian Söderquist (Berlin / New York: Walter de Gruyter, 2008) 24–37. Reprinted with permission of Verlag Walter de Gruyter.

(With Theresa M. Ellis), "Erotic Love in the Religious Existence Sphere," in *International Kierkegaard Commentary, Works of Love*, ed. Robert L. Perkins (Macon, Georgia: Mercer University Press, 1999) 339–367. Reprinted with permission of Mercer University Press.

To Harrison and Cecelia

Introduction

More than two decades ago I received an invitation from the editor of the *International Kierkegaard Commentary* series inviting me to contribute an essay to be included in a planned volume for the series. Robert Perkins was familiar with my prior work on Kant's ethics and philosophy of religion and was interested in seeing what I might bring to the understanding of ethics and sin in Kierkegaard's difficult treatise, *The Concept of Anxiety*.

Thus began an adventure that led in 1992 to the publication of my book, *Kierkegaard and Kant: The Hidden Debt*.[1] Re-examining *The Concept of Anxiety* for the first time since my undergraduate days, I was amazed by the depth of Kierkegaard's engagement with Kant's thought that I found there. This led me to a thorough comparative examination of both thinkers' writings that culminated several years later in *The Hidden Debt*. My findings were surprising. Against decades of scholarship that had identified Georg Hegel as the major philosophical influence in Kierkegaard's life, I concluded that it was Kant who furnished the main timbers of Kierkegaard's intellectual edifice. Since Kierkegaard made so few open references to Kant in his published writings (fewer than twenty versus hundreds for Hegel), I further concluded that Kierkegaard made an active effort to conceal his debt to Kant. In the intellectual world of Golden Age Denmark, where the rage was following Hegel and "going further," Kant, I argued, had become too dangerously passé to support the career of a budding young intellectual.

[1] Ronald M. Green, *Kierkegaard and Kant: The Hidden Debt* (Albany: State University of New York Press, 1992).

The deliberate hiddenness of Kierkegaard's debt to Kant was the least important of my claims, but, somewhat surprisingly to me, it dominated initial reactions to my book. Some defenders of Kierkegaard were disturbed by my perceived assault on his scholarly integrity; others believed that better explanations could be found for Kierkegaard's relative silence about Kant, such as the widespread prevalence of Kantian ideas and their assimilation into the culture of Danish intellectuals.

Fortunately, those debates about Kierkegaard's motives seem to have quieted. What remains, I believe, is a permanently changed landscape of Kierkegaard scholarship where it is now widely acknowledged that Kantian ideas played a major role in Kierkegaard's intellectual formation and shaped not only his pseudonymous and philosophical writings but also his religious discourses.

My scholarly career took a different turn from the mid-1990s on, when I served on the National Institutes of Health's Human Embryo Research Panel and was asked to help establish the NIH's Office of Genome Ethics. Bioethics became the principal focus of my work. Nevertheless, my interests in the relationship between Kierkegaard and Kant remained constant, and I found myself responding to repeated invitations to contribute an article on the topic to books, journals, or scholarly panels. The chapters in this book are a selection from more than a dozen articles or chapters I have published on this theme. Amidst the often-pressing work I was doing on genomics, cloning, and stem cells, the ties between these two great thinkers remained a source of fascination to me—a reminder of the more basic philosophic and religious interests that stimulated my career in the first place.

At first, these contributions mostly refined issues treated in *Kierkegaard and Kant*. Eventually, however, they led me down new pathways and allowed me to see connections (and disagreements) between Kant and Kierkegaard that I had not previously imagined. As I look back now on this collection of writings, I see that together they take our understanding of the Kant-Kierkegaard connection to a new level. Above all, they bring to the fore the insight that each thinker's work is an intense engagement with one question: how do time and eternity relate to one another in human life? Human moral experience, both thinkers agree, shows that we live our lives at the intersection of two competing realities—temporal experiences and timeless moral requirements. The question is, how can we unite these seemingly discordant features of our existence?

Superficially regarded, Kant and Kierkegaard offer different answers to this question. The principal matter in dispute is the need for a historical savior for the completion of the human moral project, with Kant taking a negative view of this, while Kierkegaard makes it a central claim of his authorship. But as these chapters reveal, the differences between their conclusions are less important than the very similar insights that stir their thinking. Each offers a compelling view of the human being as a creature caught up in the temporal flow of experience, unable to peer out of this flow to see where it is going, but nevertheless required to heed the dictates of an unbending call to moral perfection. For both thinkers, the choice for eternity, perceived in the fulfillment of the moral requirement, is compelling, but not necessary, while the torrent of experience provides no solid footing for the turn toward righteousness that Kierkegaard and Kant feel we must make. Hence, the resolution of this matter is always a free decision, a choice—*a leap*

The leap is differently expressed in each thinker's work. For Kant, it is a conceptual move made necessary by the limits of human cognition. Linked to reason, it modestly transcends reason in the sense that it takes us into a realm ordinarily closed to us by reason's cautions. But for Kierkegaard, the *leap* is a passionate jump beyond reason, an anguished and isolating movement into a domain that challenges all our certainties. In point of fact, the differences here are again less striking than they first appear. In important respects, both thinkers are philosophical rationalists who develop the logic and limits of human reason, but their tones and approaches are markedly different. Where Kant speaks of "the esthetic" in formal terms as a component of our cognition, Kierkegaard offers it as a style of life, an entire mode of being. Where Kant deduces free will as a presupposition of the rational moral law, Kierkegaard uses his method of indirect discourse to awaken his readers to their own responsibility for choosing which life-stance to make their own. And where Kant only rarely departs from the framework of the formal philosophical treatise, Kierkegaard uses fictional literature, poetry, and music to incarnate the ideas and choices he presents. His philosophy of existence *lives*.

The essays in this volume are presented thematically, and within each theme, in chronological order. I begin with an essay focused entirely on Kant, "The First Formulation of the Categorical Imperative as Literally a 'Legislative' Metaphor." This appeared in 1991, the fruit of nearly two decades of work on Kant's ethics. As a graduate student at Harvard, I was re-introduced to Kant by the work of the philosopher John Rawls. In powerful new ways, Rawls reconnected Kant with the contract tradition of ethics. For me, the greatest importance of Rawls's work on Kant was the new insights it provided into Kant's philosophy of religion. I later presented these insights in two books developing aspects of Kant's philosophy of religion: *Religious Reason* (1978) and *Religion and Moral Reason* (1988).[2]

---

[2] Ronald M. Green, *Religious Reason: The Rational and Moral Basis of Religious Belief* (New York: Oxford University Press, 1978); ———. *Religion and Moral Reason: A New Method for Comparative Study* (Oxford University Press, 1988).

The core insight resulting from this new approach to Kant's ethics is the idea that Kant is a philosopher of happiness. Ethical textbooks typically portray Kant as the stern philosopher of duty, the thinker who opposes the individual's pursuit of well being and who upholds as a moral ideal the cold and unsympathetic person who acts "from duty" alone. But, in fact, this is an unfair caricature of Kant's complex position. No less than the British empiricists, whose work he engaged with, Kant belonged to the Enlightenment and believed in the value of worldly fulfillment and well-being. However, unlike the British empiricists and the utilitarians who flourished in their wake, Kant realized that personal happiness could never rationally be made the supreme motivating consideration of any person's choices, what Kant called the "determining ground" (*Bestimmungsgrund*) of one's will. Doing so, he saw, was a recipe for anarchy, as each individual's unchecked private pursuits interfered with those of others and undermined the possibility of *everyone's* well being. In place of personal happiness, Kant insisted on the prior and primary obligation of obedience to the moral law, which he called the categorical imperative. Kant first discusses this imperative in his treatise, *Foundations of the Metaphysics of Morals* (*Grundlegung zur Metaphysik der Sitten*). The imperative does not rule out the search for personal happiness, but it always subordinates it to the needs of the common good. Kant believed that the categorical imperative, which he formulated in various ways, provides a complete and reliable guide to moral choice.

Over the years, readers of the *Foundations*, from academic philosophers to undergraduate students, have been intrigued by Kant's arguments. Much that Kant says about the imperative—his formulations and many of his illustrations—makes good moral sense, but some does not. Overall, the impression is that Kant was definitely onto something, but that his argument does not quite work. The conclusion is that there may be no single, rational principle or approach that can embody and express our thinking about ethics, or that such a principle is an aspiration but not a reality.

By positioning Kant's argument in a larger political and contract frame of reasoning to which I think it belongs, this first essay tries to reply to these objections and uphold Kant's strong claims for the imperative. The reader can judge how well I do. But apart from this, the essay has value for our understanding of the link between Kant and Kierkegaard. It does so by letting us see the basic conceptual problem that drove Kant to ponder the philosophy of religion and that led to his later writings—the *Critique of Practical Reason, Religion within the Limits of Reason Alone,* and *The Conflict of the Faculties*—that so influenced Kierkegaard. As beings of time and space, Kant saw that we necessarily seek the satisfaction of our needs and desires—our happiness. Yet our moral reason sometimes asks us to subordinate the pursuit of private satisfaction—and even to act in ways that terminate our ability to experience happiness at all. This all lends urgent force to the questions "Why should I be moral?" "Why should I reason in a way that leads to the defeat of one of reason's prime purposes: furthering my happiness?"

Understanding the political dimension in Kant's argument for the categorical imperative lends even greater force to these questions. Morality requires me to view my conduct, not from my personal vantage point, but from the perspective of an impartial legislator. But why should I do this? Why privilege society over myself? Why admit a societal viewpoint into the very heart of my justificatory reason? The reply that we need social order to pursue our private ends works well in most cases, but not when those ends are urgent and when we are willing to risk chaos around us to secure them. This problem evidences itself too often in the disappointing conduct of real legislators who so easily fall prey to lobbyists' generous "campaign contributions" or forsake the common good in the name of wasteful but politically beneficial pork-barrel expenditures.

For Kant, this problem was a rational one, residing at the core of our reason, which, in all its employments, relies on impartiality and objectivity as a means of adjudicating disputes. He perceived that any appeal to impartial reason to adjudicate the dispute between reason in its moral employment and reason in its prudential (private/happiness seeking) employment necessarily, and unfairly, favors moral reason. Avoiding a circular reliance on reason's own challenged modes of justification requires a solution *beyond* reason (although one indicated by reason). This led Kant to an admittedly desperate solution to a desperate problem: to propose a rationally permissible but uncertain belief (faith) in the possible reality of a realm beyond sensory experience governed by a supreme moral causality (God). By means of these concepts, the seeming conflict between reason and duty could be resolved, since in a world governed by a supreme moral causality, it is always prudent to do our duty.

The further insight that the tension between happiness and duty is ongoing for us, and the occasion for our recurrent moral failures, intensified Kant's interest in a religious solution to these problems, eventually leading him to a philosophical rediscovery of the Christian concepts of original sin and divine grace. Put simply, Kant had rediscovered, in a rigorously philosophical way, the ancient Christian teaching that human beings have a deep but freely chosen propensity to defect from the moral law.

Kant could have moved in a different direction. He could have stayed within a long tradition of rationalist theorists going back to the Stoics and argued that the contentment afforded by a good conscience is enough of a reward for duty's sacrifices. This would answer the urgent question, "Why should I be moral?" by showing that moral obedience is always rationally in our interest and requires no transcendent support. But Kant was too penetrating and too honest a thinker—Kierkegaard's favorite term for him is "*ærlige*" (honest) Kant[3]—to accept what he knew to be a question-begging response. The more basic question, and one that defies an easy answer, is why I should want to have a good conscience and make it my supreme standard of self-judgment. This question, too, Kant believed, leads us into a realm of transcendent beliefs, which we enter not by rational necessitation but by free choice and faith. As a result, Kant's penetrating ethical theory based on reason and autonomy turned backward to traditional religious teachings that had long challenged both reason and autonomy. These were the moves that Kierkegaard so appreciated when preparing for his doctoral examination and that he incorporated into his authorship almost from the start.

If this first essay affords us a better understanding of the basis of Kant's movement into religion, it also helps us appreciate the many places in Kierkegaard's writings, from *Fear and Trembling* through religious discourses like *Purity of Heart* and *Works of Love,* where he often seems to take Kantian ethics as the paradigm of what ethics means. Kierkegaard never engages in a philosophical analysis of Kant's position, but he assumes its accuracy as a description of rational morality. In *Fear and Trembling,* Johannes de Silentio boldly announces that, "the ethical is the universal" (FT 54, 68). The overt references that follow allude to Hegel and his concept of *Sittlichkeit,* but the characterization of the ethical, as the supreme *télos* and as involving no direct duties to God, involves ideas almost uniquely associated with Kant. Understanding the political metaphors inherent in Kant's imperative, which Kierkegaard may have intuited, assists us in seeing why Kierkegaard so readily assumed that ethics subordinates the individual to the common good, why he fused Kantian and Hegelian ethics, and why he pitted both of

[3] PAP VIII2 A 358 n.d.; VIII2 B 81 n.d. 1847 [JP1-649]: cf. XI A 666. 1849 [JP3-3558]; X1, 666 n.d . 1849; X2 A 501 1850.

them against the position, embodied in Abraham, that places the individual above society.

Chapter two, "Kant on Christian Love," deepens our understanding of Kant's ethics and further highlights how much Kierkegaard drew from him. Here I present Kant's penetrating interpretation of the love commandment found in Matthew 22:37–40 and in Mark 12:30–31. This commandment, Kant tells us, should not, in the first place, be understood as requiring us to cultivate various emotional states—sympathy, compassion, or love—toward the neighbor. Rather, it is directed at the will, and asks us always to subordinate our personal desires or inclinations to the principles of rational moral duty, both in its obligations and its moral ideals.

One might think that this displacement of sympathy and compassion by reasoned and willed obedience to duty diminishes the spiritual and ethical import of the commandment. But Kant argues that just the opposite is true. Duty, he tells us, has nothing to do with feelings. The individual "who believes himself kindhearted and promotes the welfare of all mankind by empty wishes" engages in a form of moral self-conceit and arrogance (LE 136). Duty, in contrast, requires us to eschew self-gratifying "talk of love and kindness" and accept a stern and unremitting discipline on the whole of life. Duty's effect is to humble us, and to make us aware of the distance between the requirement and our performance.

In his writings, Kierkegaard sometimes criticizes Kant's understanding of the categorical imperative as a self imposed law of reason. An autonomous rational standard can no more discipline us into virtue or betray the extent of our moral failures, says Kierkegaard, than governor Sancho Panza's self-administered blows to his own bottom were adequate punishment for his misconduct (PAP X2 A 396 n.d., 1850: [JP1-188]). The loftiness of the moral requirement and our infinite, sinful distance from satisfying it can only be adequately represented in the divinely revealed per-son of Christ, the self-denying, sacrificial "prototype" whose life and death make clear the full meaning of the love commandment.

Nevertheless, the disagreement between Kant and Kierkegaard here is not as great as it seems. This becomes clear when we turn to Kierkegaard's mature writings on ethics. In *Works of Love*, Kierkegaard puts forth a view of the love commandment with striking similarities to Kant's. Regardless of one's emotional proclivities, says Kierkegaard, the love commandment has imperatival force. "To love is a duty" (WL 24), says Kierkegaard, and its watchword is "You shall love!" In addition, Christian love has nothing to do with our personal emotional bonds. "Love is a matter of conscience and thus it is not a matter of drives and inclination, or a matter of feeling…." (WL 143). While erotic love and friendship are based on preference, "The Christian doctrine, on the contrary, is to love the neighbor, to love the whole human race, all people, even the enemy, and not to make exceptions, neither of preference nor of aversion" (WL 19).

In one of the most controversial sections of *Works of Love*, "The Work of Love in Recollecting One Who Is Dead," (WL 345–58) Kierkegaard offers remembrance of the dead as a paradigmatic example of Christian love. Unlike the objects of our ordinary loves, as in erotic love or friendship, the deceased cannot reciprocate our affection or regard. Hence, says Kierkegaard, there can be no question of self-love in the love we express in remembering and honoring them. Some critics of this view have traced it to Kierkegaard's fear of intimate relationships, the same aversion to attachment to flesh and blood people that earlier in his life may have driven him to break his engagement to Regine Olsen.[4] Yet, here, again, Kierkegaard is better understood as

[4] See, for example, Sylviane Agacinski, "We Are Not Sublime: Love and Sacrifice, Abraham and Ourselves," in *Kierkegaard: A Critical Reader*, ed.

evidencing his intellectual debt to Kant. In the *Foundations*, Kant offers as an illustration of the purest example of morality the individual with "little sympathy in his heart" who, nevertheless, performs his duty. Kant's point is not to elevate emotional coldness as a moral ideal, but to try to offer a "pure" example of duty without the admixture of other irrelevant and potentially distracting motivations. In exactly the same way, Kierkegaard turns to love of the dead as a controlled experiment. "When one wants to make sure that love is completely unselfish, one can of course remove every possibility of repayment. But that is exactly what is removed in the relationship to one who is dead. If love still abides, then it is truly unselfish" (WL 349).

Both Kant and Kierkegaard also view morality, whose highest expression is the love commandment, as a window on the eternal. For Kant, free obedience to the obligations and ideals of the categorical imperative takes us beyond the temporal sequence of natural causes, beyond the world of human drives and desires, and reveals our membership in a realm of intelligible beings, whose conduct is determined only by laws of reason (GR 78–82 [459–63]). Kierkegaard, too, sees the duty to love as freeing us from the ever-changing world of natural determinations. "Only when it is a duty to love," he says, "only then is love eternally secured against every change, eternally made free in blessed independence, eternally secured against despair" (WL 29).

In a discussion published subsequently to the chapter I present here, Paul Martens agrees that there are many "marked similarities" between Kant's and Kierkegaard's treatments of the love commandment. But he emphasizes what he believes are the disagreements. For Kierkegaard, says Martens, "it is impossible to fulfill the ocean of requirements which the duty to love entails." Thus, the commandment not only highlights our duty, but also

---

Jonathan Rée and Jane Chamberlin (Oxford, England: Blackwell, 1998) 129–150.

convicts us of our inevitable failure to fulfill it. This, in turn, points us to the need for a historical redeemer. "The sinner, who is unable to fulfill one's duty, can find comfort and eternal defense through Christ from the judgement of God."[5]

Martens is right to stress this disagreement. Kant and Kierkegaard part ways over the matter of the depth of our sin and the appropriate response to it. Kant will never relinquish our responsibility to undertake the autonomous renewal of our wills, and he was uncomfortable with any solution that began with another's intervention on our behalf. But this disagreement presumes a broad foundation of agreement between the two thinkers on which Kierkegaard built his own argument. In their treatments of the love commandment, both thinkers emphasize the intensity of the moral requirement; both see a proper understanding of ethics as necessarily leading to a sense of moral failure, to guilt and repentance and both see ethics as the unique entry point into the religious life.

These matters of agreement and disagreement become much clearer in my third chapter, "The Limits of the Ethical in Kierkegaard's *The Concept of Anxiety* and Kant's *Religion within the Limits of Reason Alone*." This is the essay that first spurred my engagement with the Kierkegaard-Kant relationship. Here I develop at length Kant's arguments in affirming the unyielding nature of the moral demand and its inevitable connection to human moral failure. I also trace the presence of these arguments in Kierkegaard-Vigilius Haufniensis's opaque remarks about the ideality of ethics, the infinitude of human sinfulness, and the role of Adam as the prototypical sinner. Exactly like Kant, Kierkegaard-Haufniensis rejects the traditional Christian notion of

---

[5] Paul Martens, "'You Shall Love': Kant, Kierkegaard, and the Interpretation of Matthew 22:39," in *International Kierkegaard Commentary: Works of Love*, ed. Robert L. Perkins (Macon, Georgia: Mercer University Press, 1998) 75.

inherited sin and affirms each person's culpability for sin through his or her recapitulation of Adam's choices. "To explain hereditary sin through every man's relationship to Adam and not through the individual's 'primitive' relationship to sin," Kierkegaard-Haufniensis tells us, "is to place Adam fantastically outside history and outside the human race" (CA 26–28). The willingness by Kierkegaard to depart, in the midst of a Christian-dogmatic exposition, from traditional Christian teaching and uphold the centrality of autonomy, even while affirming the reality of sin, is a vivid illustration of how much he absorbed from Kant's thinking. This chapter, which was written before I began the research for *The Hidden Debt,* closes with a question and remark that, in retrospect, I find amusing. "Did Kierkegaard, then, actually study the *Religion* and draw upon it in writing *The Concept of Anxiety?* This question has been left for last because presently I am unable to answer it." Today I can say with confidence that Kierkegaard not only read the *Religion,* he engaged in an intense dialogue with its views on ethics, sin, the Bible, and father Abraham.[6]

Chapter four, "Kant and Kierkegaard on the Need for a Historical Faith: An Imaginary Dialogue," was originally inspired by an invitation to participate at a conference organized by D.Z. Phillips at Claremont Graduate University in 1998 on Kant and Kierkegaard's philosophy of religion. The graduate school context prompted me to try a somewhat lighter touch, and my concurrent work on genomics spurred the idea of imaginarily cloning these two great thinkers in order to facilitate a conversation between them. (I am aware, of course, that a cloned Kant and Kierkegaard would not have the contents of either man's intellect or memory, so I enlisted computer technology for that purpose.) The resulting dialogue, while humorous, has a serious purpose. It attempts to answer the question of what Kant and Kierkegaard would each

[6] For references to these passages, see Ronald M. Green, *Kierkegaard and Kant: The Hidden Debt,* 17–18.

have thought about the other's work at the crucial points where it connects to their own. In some ways, this chapter recapitulates material in the previous and following one, but it places these issues in a *lived* context appropriate to both thinkers. One additional comment: When I first delivered this paper at Claremont, it had Kant and Kierkegaard ordering Tuborg beer from the waiter. A member of the audience reminded me that Kant was a wine drinker, and would not order beer, which, in the *Anthropology*, he accused of rendering people withdrawn and less talkative.[7] In this version of the dialogue, our heroes order California Chardonnay.

Chapter five, "A Debt both Obscure and Enormous," marks the most recent point on the trajectory that began with my *IKC* essay on *The Concept of Anxiety*. It was invited for a volume on Kierkegaard and his German philosophical contemporaries edited by Jon Stewart at the Søren Kierkegaard Research Centre in Copenhagen.[8] Like all the chapters in Stewart's volume, it was meant to be a comprehensive analysis of the scope of Kant's influence on Kierkegaard, and in an accompanying section not reprinted here, it contained a listing of the writings by Kant and materials dealing with Kant that were in Kierkegaard's library at the time of his death, as well as more recent secondary treatments of the Kant-Kierkegaard relationship. While this chapter repeats several key arguments in the two preceding ones, it has independent value, I think, as a concise and thorough treatment of the issues central to Kierkegaard's engagement with Kant on the matters of ethics, sin, and the role of historical knowledge in our redemption.

---

[7] Robert B. Louden, ed., *Anthropology from a Pragmatic Point of View* (Cambridge: Cambridge University Press, 2006) 63.

[8] Jon Stewart, ed., *Kierkegaard and His German Contemporaries*, ed. Tome 1: Philosophy, in the series "Kierkegaard Research: Sources, Reception and Resources" (London: Ashgate, 2007) 179–210.

The chapter also brings to the fore in a new way the specific *epistemological* influence of Kant on Kierkegaard. Kant's fame as a philosopher rests, in part, on his penetrating refutation in the first *Critique* of a centuries-long tradition of rational theology based on the ontological argument for God's existence. That argument includes existence—being—among the list of excellent qualities (predicates) that an object might possess. Since God is thought of as the most perfect being, he could not lack this predicate. It follows logically that he must possess being, that is, he must exist. Kant's rejection of this argument rests on the observation that existence is not a quality that we predicate of an object. Rather, it is the affirmation that the object, with all its predicates (or lack thereof), is given to us in experience. "A hundred real thalers do not contain the least coin more than a hundred possible thalers," said Kant in his famous quip. "My financial position is, however, affected very differently by a hundred real thalers than it is by the mere concept of them...." (KRV 505).

Kierkegaard's appreciation of Kant's argument permeates the *Fragments* (and to a lesser extent, the *Postscript*). It appears in his distinction between "ideal being" and "factual being," "concept existence" and "real existence" (PF 39ff.). And it re-appears at the level of humor, with Kierkegaard observing that "A fly, when it is, has just as much being as the god" (PF 41n.). We can ask, however, why Kierkegaard-Climacus devotes so much space in the *Fragments* to the discussion of these epistemological issues, when the work itself, in the words of its epigraph, is focused on the question of whether "an eternal happiness can be built on historical knowledge?" (PF 1).

The answer to this question, I maintain, begins with Kierkegaard's complex encounter with Kant's discussion of human sin. Kierkegaard knew that Kant, after having developed the problem of the intractability of sin in the *Religion*, sought a solution to the problem that would be consistent with his stress on human autonomy and reason. That solution involved repentance and the trust that one's penitent rededication to righteousness was, in itself, a sign of a graciously restored and renewed will. Kant saw a role for divine grace in this process, but he set aside as not being helpful to our moral salvation questions regarding its mode of operation. As to the certainty of our reception of grace and, thus, the reliability of our renewed moral disposition, Kant says there is "no other way we can conceive the decrees of a holy and benevolent law-giver," than to assume that he will accept and sustain our penitent spirit (CF 83). In essence, Kant offered an entirely conceptual solution to our problem of defective moral willing.[9] We overcome our past moral mischoices merely by thinking our way out of them: by basing our hope for grace on the reasonable possibility of its availability to one who yearns for it.

Kierkegaard will have none of this. In the *Concept of Anxiety*, he had closely followed Kant in affirming that our original choice of evil is always a logically unnecessitated event in time: an inexplicable free choice to defect from the moral law. He also followed Kant in seeing that this defection leads to unremitting suspicion about the integrity of our will and to a state of moral despair. But if evil is a free and logically unnecessary event of this degree of seriousness, its remedy cannot be a reasonable presumption of the possibility of divine support. Such presumption would not only limit God's freedom to act (or not act); it would also reduce the gravity of our initial free choice. Everything would become part of a logically necessary process leading to our redemption.

With Kant—but now against Kant—Kierkegaard affirms the reality of human freedom and the importance of time in relation to

---

[9] We have here, I believe, another Kantian transcendental deduction (see chapter seven), that is, a "not without" argument to the effect that our undeniable experience of repentance and our will to reform logically presuppose their possibility.

it, since time is freedom's sphere of operation. Just as we are not necessitated to sin, so God cannot in any way be necessitated to save us. Both events belong to the sphere of time and being. Although the necessary always *is* (PF 74), events in time can either be or not be. God could have withheld forgiveness, and human sin could have eventuated in the unsparing punishment that it invites. The actual existence of anything, including an act in history, is therefore never necessary. The crucial question becomes whether God *has* entered time, whether he *has* manifested his forgiveness in our experience. This is a question of factual, historical being, not logic—of "real existence," not "concept existence." Against this background, the *Fragments* affirms the importance of God's initiative in time as the active savior-teacher, and not merely as an occasion for the student's eternal (and necessary) self-recollection. To the book's opening question, "Can an eternal happiness be built on historical knowledge?" Kierkegaard-Climacus replies that it must be. If Kant is right about human freedom and about the non-necessity of events in time, then history, for both man and God, becomes a crucial site for the working out of our ethical-religious destiny.

The *Fragments* is Kierkegaard's reminder to Kant that for human beings, decisions in time—our decisions and God's decisions—shape our eternal destiny. It is interesting that nearly a century later, the Jewish philosopher Emmanuel Levinas, who was influenced by both Kant and Kierkegaard, when speaking about the possibility of divine punishment, expresses a similar point:

> God's goodness dialectically brings forth something like God's wickedness. This is no more difficult to admit than many Christian mysteries. The idea that divine patience might have come to an end, and that there are sins committed, is the condition for the respect given by God to the fully responsible man. Without this finitude to divine patience, man's freedom would be only

provisional and derisory, and history just a game. We must recognize man's coming of age.[10]]

In *The Sickness unto Death* Kierkegaard anticipates Levinas's insight by emphasizing the decisive importance of the Christ event: "When God lets himself be born and become man, this is not an idle caprice, some fancy he hits upon just to be doing something, perhaps to put an end to the boredom.... No, when God does this, then this fact is the earnestness of existence" (SUD 130).

Chapter six, "Developing Fear and Trembling," is the culmination of work that began with my research for *The Hidden Debt*, but which was developed further in a later series of articles. One of these, "Enough Is Enough! *Fear and Trembling* is *Not* about Ethics,"[11] challenged the decades of scholarship that had interpreted Kierkegaard-Silentio's book as a treatise on religious ethics, specifically, a defense of a radical, divine-command ethical position. Against this, I argued that *Fear and Trembling* is principally a work of classical Christian soteriology that argues for our need for faith in the possibility of divine grace as mediated through Jesus Christ. Underlying my argument was an awareness of how much Kant had engaged with the figure of Abraham, both in the *Religion* and *The Conflict of the Faculties,* not primarily in terms of Abraham's moral conduct, but in the context of Kant's sustained denial of any role for historical or direct revelation in our moral salvation (with Abraham and Isaac symbolizing, among other things, the relationship between God and Christ). Equally important are the many signs placing *Fear and Trembling* within the events of Kierkegaard's life, notably the familial tradition of

[10] Emmanuel Levinas, "Simone Weil against the Bible (1952)," in *Difficult Freedom: Essays on Freedom* (Baltimore: The Johns Hopkins University Press, 1990) 139.

[11] Ronald M. Green, "Enough Is Enough! *Fear and Trembling* is *Not* about Ethics." *Journal of Religious Ethics* 21, no. 2 (Fall 1993): 191–209.

sin that caused him to break his engagement to Regine Olsen. On this reading, Søren is not Abraham who must emotionally kill his beloved (Regine) in order to obey God's command to lead a religious life; rather, he is the son imperiled by his father's deeds, both of whom might be saved by God's gracious intervention.

While not retreating from my claim that sin and grace are the central themes of *Fear and Trembling*, I admit here the reasonability of interpreting the text at many other levels. Certainly, *Fear and Trembling* anticipates the call to radical Christian engagement found in many of Kierkegaard's later writings. It also argues for going beyond Kant and Hegel to a revealed ethics of divine command, which we know that Kierkegaard found attractive.[12] Most importantly, this chapter reflects the understanding that Kierkegaard's writings are extraordinarily complex, integrating many different themes, approaches, and commitments within what often seem to be relatively simple surface texts. Indeed, this understanding is supported by the awareness of how effectively Kierkegaard wove Kant's philosophy into his authorship without revealing either Kant's presence or their shared philosophical concerns.

Chapter seven, "*Fear and Trembling*: A Jewish Appreciation," represents a first sortie into an issue to which I hope to return in the future: Kierkegaard's place in the landscape of nineteenth-century Danish and European attitudes to the Jews. Kierkegaard's published and unpublished writings show that, like so many of his contemporaries, he harbored stock anti-Semitic sentiments. His attitude toward Jews was not improved by his encounter with Meir Goldschmidt, the Jewish editor of *Corsair*, the satirical journal that did such damage to Kierkegaard through its vulgar ridicule of him. (It is noteworthy that Goldschmidt's novel *A Jew* [*En Jode*]

[12] For a defense of Kierkegaard as a divine command ethicist, see C. Stephen Evans, *Kierkegaard's Ethic of Love: Divine Commands and Moral Obligations* (New York: Oxford University Press, 2004).

offers its own picture of the prejudices of Danish Christians like Kierkegaard.[13]) Nevertheless, I argue here that by affirming the centrality of Abraham as the father of Jewish and Christian faith, *Fear and Trembling* remains a philo-Semitic work, exhibiting none of the profound anti-Semitism found in the writings of Hegel. If Hegel, in this respect, can be seen as pointing ahead to the Aryan Christianity of the Nazi period that would sever the Hebrew Bible from the canon,[14] Kierkegaard remains on solidly biblical ground. In the future I hope to return to Kierkegaard's relationship to Judaism, which has already been treated by several scholars.[15] In this theme, I believe, we have a sensitive measure of the inner integrity and ethical directionality of Kierkegaard's Christianity. Just as feminist scholars have plumbed Kierkegaard's attitudes toward women as a measure of his thought,[16] so scholars interested in Judaism (and other minority traditions) should bring their interests and questions to his work.

Chapter eight, "Kierkegaard's Great Critique: *Either/Or* as a Kantian Transcendental Deduction," exposes his debt to Kant's philosophical method. Invited by the *International Kierkegaard Commentary* series again to examine Kant's pertinence to a work by Kierkegaard (this time, *Either/Or II*), I found a major methodological parallel between Kant's employment of transcendental

---

[13] Meir Goldschmidt, *A Jew (En Jode)*, Kenneth H. Ober, trans. (New York: Garland, 1990).

[14] Susannah Heschel, *The Aryan Jesus: Christian Theologians and the Bible in Nazi Germany* (Princeton: Princeton University Press, 2008).

[15] Bruce H. Kirmmse, "Kierkegaard, Jews, and Judaism," *Kierkegaardiana* 17 (1994) 83–97; Jennifer Goolsby Pouya, "Kierkegaard and the Jewish Shadow" in *Kierkegaard and Religious Pluralism*, ed. Andrew J. Burgess, papers of the AAR Kierkegaard, Religion and Culture Group and Kierkegaard Society, American Academy of Religion 2007 Annual Meeting, San Diego (Eugene, Oregon: Wipf and Stock Publishers, 2007) 31–53.

[16] See, for example, *Feminist Interpretations of Søren Kierkegaard: Re-Reading the Canon*, eds. Céline Léon and Sylvia Walsh (University Park: Pennsylvania State University Press, 1997).

deductions and the Judge's penetrating arguments against the position of the young esthete. The transcendental deduction was perhaps Kant's major philosophical tool. It involves the use of a "not without" argument to show that a given experience rests necessarily on certain conceptual presuppositions, that the experience "could not" take place "without" these concepts. Kant uses such deductions to ground the a priori elements of our cognition, to demonstrate the necessity of our assumption of freedom, and, also, I believe, to justify the possibility of grace in our moral redemption.

In my research, I was interested to learn that the idea of a transcendental deduction, which in the first *Critique* is redolent of legal terminology, actually goes back to a long tradition of litigation over rights of possession. In this chapter, I argue that Judge William fits comfortably, as judge-litigant, into this tradition. Arguing against the esthete's claim to possess real apprehension of the experience of "first love," he deploys a complex "not without" argument to justify the necessity of ethical resolve to understand and fulfill our romantic-emotional life. Scholars continue to dispute how much of the Judge's ethical position as outlined in *Either/Or* II is traceable to Kant. But this chapter indicates that the very method used by the Judge may be influenced by Kant's philosophy.

In chapter eight, I begin to develop my sense that *Either/Or* is built on the main lines of Kant's philosophy and is meant to be a preparatory introduction or "propaedeutic" to Kierkegaard's entire authorship. Chapter nine, "*Either/Or*: Kierkegaard's Overture," continues my engagement with this work and develops *Either/Or*'s foundational role using a new metaphor. This was written for the 2007 Søren Kierkegaard Centre's annual summer research conference, which was focused on the first volume of *Either/Or*. I had not previously been a great fan of *Either/Or I*, regarding it as a wordy and somewhat immature first effort by Kierkegaard, more expressive of his literary than his philosophical interests. My more recent use of the volume in teaching has softened that judgment, since I find that college students are more stimulated by the esthetic style of life there than they are by the other life-stages presented in Kierkegaard's writings.

In the course of teaching *Either/Or I*, I frequently choose to show a video performance of Mozart's *Don Giovanni* to help students better understand the essay on "The Musical-Erotic." This led me to pay more attention to what Kierkegaard-Eremita has to say about operas and operatic overtures in that essay. Spurred by the insight that there are often arguments within arguments in Kierkegaard's writings, the thought occurred to me that maybe Kierkegaard was saying something about his own work (his own *opera*). This chapter is the result, arguing that *Either/Or* in its entirety is the overture to Kierkegaard's authorship.

What is most noteworthy in this piece, I believe, is the importance it lends to another literary component of *Either/Or I*: the essay on the "The Tragic in Ancient Drama Reflected in the Tragic in Modern Drama." Here I argue that Kierkegaard-Eremita uses the *Antigone* to prefigure the deep family drama and preoccupation with sin that will reemerge in *The Concept of Anxiety* and *Fear and Trembling*. Indeed, Antigone herself prefigures Isaac in the subsequent work. True to Kierkegaard-Eremita's ideal of an overture, the ideas presented here are only a suggestive motif, but also true to his concept of an overture, this motif is highly revealing. It indicates a deep and abiding theme of Kierkegaard's authorship as a whole. Most interesting is the possibility that the sexual ideas present in the figure of Antigone will help us better understand Kierkegaard-Climacus's opaque discussions of hereditary sin in *The Concept of Anxiety*. More than any other figure in Kierkegaard's writings, Antigone (in both her ancient and modern dress), lets us see how we can become actively involved in and responsible for the misdeeds of our forebears, including their sexual misdeeds. Just as the violin notes bearing tones of death and anxiety in Mozart's overture lend great meaning to the later scenes where Don Giovanni meets his fate, so this hint of the theme of hereditary and sexual sin in *Either/Or* illuminates the reappearance of these ideas throughout Kierkegaard's *opera.*

The final chapter of this book, "Erotic Love in the Religious Existence-Sphere," is the fruit of a collaboration between myself and my undergraduate Dartmouth honors student Theresa Ellis. With my encouragement and advice, Theresa had examined the concepts of romantic love in Kierkegaard's writings. In her thesis, Theresa identified the six features of romantic love discussed here. Foremost among these is the sense of eternity that marks love, especially first love. In this chapter, we join these insights to an examination of how love evolves as it expresses itself in each of the different existence-spheres. This examination leads to the conclusion that romantic (and even sexual) love might not vanish in the religious stage, but rather be radically transformed, fulfilled, and ennobled by the awareness of sin and the presence of God's sustaining grace. Hence, this chapter returns us to the themes of sin and grace, time and eternity that I believe form the center of Kierkegaard's writings. In romantic love, as in all human ethical experience, we are creatures who perceive and work out our eternal meaning *within* the temporal flow of experience.

This chapter also shows us how different Kierkegaard can be from Kant. Kant has no philosophy of human emotional and erotic love. He shies away from such intimate engagement with the lived experience of human beings. Kierkegaard, in contrast, revels in bringing his concepts into our lives. It is not that Kierkegaard rejected Kant's key philosophical insights. On the contrary, he used them and carried them further. But he always sought to incarnate them in the material of human experience.

Understanding these similarities and differences between Kant and Kierkegaard sheds new light on the trajectory of philosophical development between the nineteenth and twentieth centuries. While Kant's basic epistemological and ethical ideas serve as the foundation of Kierkegaard's work, Kierkegaard gives these ideas new form and presents them in new expressions, illustrations, and with new substance. He uses human relationships, especially the experience of romantic love, as a vehicle for understanding the significance of our location in time and eternity. The result, in many respects, is a new philosophy. Here, Kierkegaard serves as the crucial transitional link between Kant's formal philosophy of freedom and existence and the very different existential philosophies and literatures developed by writers like Martin Buber, Paul Tillich, Jean Paul Sartre, Simone de Beauvoir, and Albert Camus.

As a New Englander, I never cease to admire the beautiful white Congregationalist churches of our region. As I point out in chapter eight, the builders of these churches took over the Christopher Wren designs they had learned from England, but erected the structures in wood—the most available material of their region. To protect the buildings from the weather, they again used material at hand and painted them white. The result is something surprisingly different from the English exemplars. The heavy and reserved English buildings now soar heavenward and radiate the light of their Christian faith. Much the same is true, I think, of what Kierkegaard did with Kant's writings. In his hands, Kant's weighty philosophy *lives.* To those Kierkegaardians who still believe that I impugned the master by saying that he borrowed heavily (and often surreptitiously) from Kant, I can only reaffirm my conviction that Kierkegaard's borrowing soars heavenward.

1

The First Formulation of the Categorical Imperative as Literally a "Legislative" Metaphor

Imagine the following situation: A senior member of Congress is being pressured by a lobbyist to support a bill that the lawmaker, on the basis of all his past experience, knows to face insurmountable opposition. After repeated efforts to state the problem to the lobbyist, the legislator finally says in desperation: "Listen, my friend, can you seriously propose this bill to me?"

The meaning of this question is clear. The legislator is not asking whether the lobbyist favors the bill, nor whether he wants to see it become law, since both facts are presumed by the situation. Rather, he is asking the lobbyist whether he can reasonably believe that his proposal will garner sufficient support—will be voted for by enough lawmakers—to pass from bill to law. The fact that it cannot do so is advanced as a reason for not proposing it in the first place.

As this imaginary encounter shows, our ability to propose policies sometimes depends on their suitability for surviving a legislative procedure where the support of other persons is required. As obvious as this understanding is, it is remarkable that it has virtually never been applied to the test of policies (maxims) implicit in Kant's first formulation of the categorical imperative. That formulation, "Act only according to that maxim by which you can at the same time will that it become a universal law" ("Handle nur nach derjenigen Maxime, durch die du zugleich wollen kannst, dass sie ein allgemeines Gesetz werde" GR [421]), has usually been interpreted as involving only the test of rational consistency or universalizability. Although various interpretations of what this means have been put forward, all involve the idea that the choosing agent need only be able to will the maxim as being *acted on* by all other similarly situated rational persons.[1] This universalized maxim is held to be acceptable if it is logically conceivable and if the agent would personally be willing to see it widely practiced. Altogether lacking in this approach, however, is any sense that the test involved here has a literally legislative dimension, that the acceptability of the universally extended practice might depend not just on the wishes of the proposing

---

[1] A list of references to all those who have interpreted the first formulation as involving only universalizability (or rational consistency or generalization, as it is variously called) would include the great majority of Kant's commentators and critics. More remarkable than the number of commentators who have read Kant this way, however, is the unreflective way in which even sophisticated commentators slip into this interpretation. Several representative quotations illustrate what I mean:

"I shall assume that what Kant means by saying that we cannot will that a maxim should become universal law is that, if it were within our power to bring about a state of affairs in which everyone acted on our maxim, we could not bring ourselves to do it."—Jonathan Harrison, "Kant's Examples of the First Formulation of the Categorical Imperative," *The Philosophical Quarterly* 7 (1957): 52.

"The principle we have just quoted (the first formulation), means that people should only adopt as rules of living for themselves rules they can will that should always be followed by everyone.—H. B. Acton, *Kant's Moral Philosophy* (London: Macmillan and Co., 1970), 21.

"'Führe nur generalisierbare Handlungen aus!' ist keine Interpretation. Es ist nur eine kürzere Fassung des kategorischen Imperativs," ("'Perform only generalizable actions' is no interpretation. It is only a short version of the categorical imperative.")—Manfred Moritz, "Über einige formales Strukturen des Kategorischen Imperativs," *Kant-Studien* 65. Jahrgang-Sonderheft, Teil I (Akten des 4. Internationalen Kant-Kongresses Kant-Studien-Sonderheft, Hrsg. (Berlin/New York: Gerhard Funke und Joachim Kopper, 1974–75): 201.

agent, but also on the willingness of *other* possible agents to vote for the practice and allow it to become a law for their community.

In the remarks that follow, I want to contend that the first formulation is meant literally to be a legislative metaphor in this sense. I also want to argue that this legislatability understanding of the test is an entirely natural reading of Kant's statements of the first formulation of the categorical imperative. I do not wish to maintain that Kant always had this understanding clearly in mind. As we shall see, on at least one occasion he speaks in terms reminiscent of the universalizability view with its referent to the individual's willing alone. Nevertheless, I hope to show that when the matter is examined without prejudice, there is textual evidence to suggest that Kant held a literally legislative conception of the imperative he was proposing. I also wish to maintain that understanding the categorical imperative in this way frees it from some of the most serious objections that have been raised against it, and renders it precisely the supreme guide to moral judgment Kant says it is.

Before turning to the texts themselves, let me state clearly how I believe the first formulation should read. This is needed because, as presented, the first formulation is elliptical. Thus, even if we concede that Kant is maintaining that maxims be capable of literally being passed into law (*Gesetz*), it is not clear just what kind of legislative process he has in mind. Specifically, who are the legislators who must vote for a maxim? What community do they represent, and how many of them must support a proposal for it to become the established and binding regulation of their community?

At the risk of being too brief here—the subsequent theoretical discussion will clarify my meaning—I wish to suggest that the answer to all these questions can be found in the single adjective "universal" (*allgemein*) which qualifies law in every statement of the first formulation. Commonly, this adjective is taken to describe the scope and object of the possible legislation, its universal applicability. A universal law is one governing *everyone's* conduct. However, if a democratic polity is assumed, the adjective "universal" can also be understood to describe the source and origin of the legislation, and thus as describing the community of persons who have brought it into being. A "universal law," therefore, may not only *apply* to all but also be the law *of* all, just as French law both applies to all French citizens and is regarded as having been made or accepted by them. Understood this way, Kant's phrase "universal law" provides answers to all the legislative questions just posed. The law which maxims must be capable of becoming is a law for the whole community of rational beings (all those beings capable of understanding and respecting law). It not only applies to all these persons, but it is *their* law in the sense that all have approved its passage from proposal to legislation and all have freely accepted it as binding on their conduct.

The idea of a policy being freely accepted as binding by all rational persons is undoubtedly very complex. In the theoretical discussion ahead, I shall look at the details of this more closely and we shall see that there is nothing either fantastic or unworkable about the idea. For now, however, I do not want to defend the conceptual merit of this notion. My aim is to suggest it as an amplification of the reading of Kant I have proposed. Thus, whenever Kant states the first formulation, I am suggesting that he is elliptically stating the following, more complex formula: *"Act only on that maxim which you are rationally able to will as being approved, accepted and acknowledged by all other rational agents as a law governing everyone's conduct."*

As substantial as this amplification of the first formulation may seem, I believe that a straightforward reading of the first formulation suggests it as naturally as any other. True, the first formulation asks whether *you* can will a maxim to be a universal law, and this may appear vaguely to support the universalizability interpretation with its focus only on the coherent volition of the individual agent. But the individual's ability to will is at issue in either interpretation. The question is which further considerations determine the individual's ability to will. Beyond the logical coherence of what I propose, is it a matter only of what I am willing to put up with, or does my ability to will also depend on others' acceptance of my proposal? As stated, the first formulation provides no answer to these questions.

In view of the genuine ambiguity of the formulation and its susceptibility to this legislative interpretation, why has the universalizability reading always seemed so natural? The answer, I think, is not hard to find. In the first section of the *Grundlegung*, which deals with "common rational knowledge" and is not, therefore, philosophically precise, Kant offers a preliminary statement of what will become the first formulation. That statement, "I should never act in such a way that I could not also will that my maxim should be a universal law" ("ich soll niemals anders verfahren, als so, dass ich auch wollen konne, meine Maxime soll ein allgemeines Gesetz werden" (GR 18 [402]), is immediately followed by an application strongly suggestive of the universalizability interpretation. Considering the question of whether a deceitful promise is consistent with duty, Kant proposes that "the shortest but most infallible way" (die allerkürzeste und doch untrügliche Art) to find the answer to this question is to ask: "Would I be content (würde ich wohl damit zufrieden sein) that my maxim (of extricating myself from difficulty by a false promise) should hold as a universal law for myself and for others?" (GR 19 [403]).

Clearly, the language used here by Kant strongly suggests the test of maxims associated with the universalizability view. Everything seems to depend on the agent's own *willingness* to put up with his or her maxim extended to others. There is no suggestion that the agent need also consider the *acceptability* of this maxim to others. Bearing in mind that this application follows Kant's preliminary mention of morality's supreme rule, we should not be surprised if this discussion decisively shapes readers' approaches to the later explicit discussion of the imperative itself.

Even as we acknowledge the force of this preliminary discussion however, it must be noted that this is the *only* instance in the *Grundlegung* where Kant uses such narrowly self-referential language in specifying the test of maxims the categorical imperative involves. True, in the well-known "law of nature" variant of the first formulation, Kant places particular emphasis on the agent's will: "Act as though the maxim of your action were by your will (durch deinen Willen), to become a universal law of nature" (GR 39 [421]). This may suggest that the agent's own consistent willing is sufficient to establish a universal law. Nevertheless, the natural law one imaginatively wills into being here is only tentatively established and must still be assessed. The critical question is what final condition governs its acceptability. Does it depend merely on the considered wishes of the willing agent or on the corresponding wishes of all other agents as well? The "law of nature" variant is silent on this matter.

If we put aside these ambiguous law-of-nature passages, we can say that outside of the suggestion in GR [403], Kant nowhere else in the *Grundlegung* uses any self-referential language in connection with the first formulation. Elsewhere, the broader terminology "can (or able to) will ... as universal law" prevails (GR 41, 49, 55 [424, 431, 436, 437]), and such terminology also permits a legislative reading. Furthermore, even in the decisive preliminary application (in GR 19 [403]), Kant's description of the test as "the shortest but most infallible way" of evaluating a maxim suggests that he is not displaying the full import or application of the principle he is proposing but merely presenting a useful shortcut application that will work in cases like this. If we bear in mind that even where the more complex legislative test is presupposed, one's own considered refusal to see one's self-serving maxim universalized counts decisively against its ever becoming law. Thus, it may be that Kant has in mind here only a quick way of screening maxims about which we have strong intuitive doubts. Alternatively—and I lean toward this view—it may be that Kant is merely a bit confused about the complexities of the test he is proposing and is not here fully aware, as we shall see, of how much a universalizability and legislative test can diverge. Whichever the case, this hint of the universalizability view is only that: a hint. It is followed up virtually nowhere in any ensuing discussions of the imperative in the *Grundlegung* or elsewhere.[2] Nevertheless, we can understand its decisive impact. At this early point in the *Grundlegung,* generations of readers have been switched onto a track of interpretation, which, given the ambiguity of the first formulation, is thereafter nearly impossible to escape.

Interestingly, even leading translators of Kant may have allowed the early 402–403 discussion to shape their judgment. Thus, both Lewis White Beck and H.J. Paton insert the *should* (*soll*) of the very preliminary formulation (*soll ... werden*) into their translations of the more precise statement of the imperative at 421, even though this word is lacking in the latter passage where the phrase *allgemeines Gesetz werde* is used instead. Beck, for example, translates the imperative as *Act only according to that maxim by which you can at the same time will that it should become a universal*

[2] In one other instance, in a passage in the *Critique of Practical Reason* (KPV 72), Kant explicates the law of nature variant of the first formula in the same subjective-volitional terms. He observes that people do conform to rational morality by asking the question: Would a person assent of his own free will to being a member of an order of things where personally advantageous deception is practiced? Once again, however, Kant is not offering a rigorous philosophical test but a common-sense formula, and the maxim at issue reflects conduct bound to fail on any test.

law (GR 89) This translation is not strictly incorrect since the simple *werde* of the German can be rendered as the Anglo-English subjunctive form *should become,* in the sense of *might become.* But *should become* (especially in American English) also has the meaning of *ought to become,* an expression which, in the context of the statement of the categorical imperative strongly points to the self-referential desirability test of maxims associated with the universalizability view. Despite Beck and Paton, however, there is no *soll* or construable *ought to* meaning in the precise formulation of the imperative. The test is not whether I think a maxim *should* or *ought to* become a universal law or, what is the same thing in this context, whether I would choose that it become so, but whether it *might become* such a law. Is it possible that the preliminary expression's use of *Soll ... werden* (should or ought to become) along with the subsequent example has led these translators to skew the later 421 passage in a way that emphasizes the role of the agent's preferences in this test and that has, in turn, influenced subsequent readers of Kant via these important translations?[3]

To this point my argument for a political interpretation has, at best, been negative. I have only indicated that we are not compelled to accept a universalizability reading of the imperative on the basis of one limited and unrepresentative passage. Is there a more positive argument on behalf of the legislative view? Here I must be less than decisive. While there is some textual evidence supporting this reading of the imperative, no single passage exists where Kant states that the principle should be understood this way. This means that in order to ground this interpretation more firmly we are led to the kind of independent theoretical analysis in

[3] I note that even as sophisticated an interpreter of Kant as Onora Nell employs Paton and Beck's translation of the 421 passage as the basis of her account of the imperative. See her *Acting on Principle* (Columbia University Press, 1975) 1, 32.

which I will engage in the second half of this discussion. More on that shortly, but first the textual evidence.

When we turn to Kant's explicit discussions of the first formulation, clear signs of a legislative interpretation are lacking, but so too is evidence for the universalizability view. Kant immediately deals with certain maxims, including the making of false promises, the thought of whose universal extension as a practice seems to lead to logical incoherence. But, if Kant is right, such universalized maxims are self-contradictory and cannot be *thought* of as universal law on any construal of this test (whether as universalized practices or as universally accepted laws). But Kant then quickly moves beyond these examples to others, including the so-called imperfect duties where, as Paton and others have noted,[4] the emphasis is not on simple contradiction in the agent's volition so much as contradiction in terms of some kind of teleologically conceived natural order. Repeatedly, the question is not whether the agent is willing to put up with others acting on the maxim, but whether the resulting practices might contradict certain recognized, normative purposes of nature that no rational person dare contravene. Thus, in discussing suicide, self-neglect, and the refusal to aid others, Kant seems to consider these maxims unacceptable because the resulting laws oppose the natural necessity of self-preservation, self-development, and mutual aid.

Kant's arguments here have been criticized in various ways,[5] but it is worth noting that these arguments have little relation to

---

[4] H.J. Paton, *The Categorical Imperative* (London: Hutchinson University Library, 1947) 153–55. See also Keith Ward, *The Development of Kant's View of Ethics* (Oxford, England: Basil Blackwell, 1972) 107–13; and Bruce Aune, *Kant's Theory of Morals* (Princeton: Princeton University Press, 1979) 59f.

[5] J. Kemp, in his article, "Kant's Examples of the Categorical Imperative," *The Philosophical Quarterly* 8 (1958): 63–71 (at page 65) points out that even on a teleological or natural-law reading, Kant's approach to the matter of suicide seems confused. Elsewhere, I argue that Kant's treatment of suicide and most

the universalizability view with its reference to individual rational volitions. Furthermore, going beyond Paton and others, I would point out that this teleological or natural-law reading is fully compatible with the political interpretation I am proposing. For if one believes—as Kant surely did—that there is a range and hierarchy of human purposes discernible in nature, then it will also be true that, even where idiosyncratic individuals might rationally permit violations of these purposes, such maxims would probably not be accepted as law by the entire community of rational agents. In the case of suicide, for example, the adoption of a maxim permitting self-destruction whenever life's suffering exceeds its pleasures might be rationally willable for a despairing individual who believes life is meaningless, but from the standpoint of the entire community of human beings, this maxim could lead over time to the extinction of the human life and to irreversible loss for everyone, whatever value they placed on life. In other words disteleological practices that might be rationally acceptable to an individual become far less so when viewed from the perspective of an ongoing human community. If this is correct, Kant's teleological excursions may be seen to fit within a more basic legislative conception of the categorical imperative. The fact that Kant does not content himself with the statement of a classical natural law or teleological view but strains to provide a more basic rational test of maxims suggests that he has this larger conception in mind for evaluating and ordering the natural purposes he discerns.

Positive evidence for this legislative interpretation lies outside Kant's immediate discussion of the first formulation. It is found first of all elsewhere in his writings, and second, in his

---

sexual sins is an ill-fitting part of his argument in the *Foundations*, attributable not to a careful, rational analysis of the issues but largely the result of his received Christian-ethical views. See my "What Does it Mean to Use Someone as 'A Means Only': Rereading Kant," *Kennedy Institute of Ethics Journal* 11, no. 3 (2001): 249–263.

further efforts to formulate the imperative in the *Grundlegung.* As early as the *Critique of Pure Reason,* for example, Kant defines reason in terms of the idea of universal free assent:

> Reason depends on this freedom for its very existence. For reason has no dictatorial authority; its verdict is always simply the agreement of free citizens, of whom each one must be permitted to express, without let or hindrance, his objections or even his veto. (KRV 593)

If we keep in mind Kant's repeated insistence that morality is fully an expression of rationality, it is fair to presume that the practical rational test he would impose on maxims conforms to this general description of reason. Hence, a maxim's moral acceptability involves passing a legislative test: its receiving the vote—and avoiding the veto—of every rational person.

That Kant's ethical theory is also thoroughly permeated by legislative concepts and notions of universal assent to principles is widely acknowledged.[6] The precritical *Dreams of a Spirit-Seer* (1776), for example, reflects Kant's great debt to Rousseau's conception of the General Will in defining morality as involving the "felt dependence of the private will upon the Universal Will" (D 51]). More than two decades later, in the *Metaphysics of Morals,* Kant defines an action as right "if it can coexist with everyone's freedom in accordance with a universal law" or if "the freedom of choice of each can coexist with everyone's freedom in accordance with a universal law" (MS 56). Here and elsewhere it is the idea of an uncoerced harmony (*Einstimmigkeit*) of wills in a community under laws that marks Kant's thinking. But clearly, such harmony would not result from any individual's willingness to have his or her maxim acted on by others, as the universalizability view requires, since this would produce at best uniformity of behavior,

---

[6] The legislative dimensions of Kant's thinking have been especially illuminated by Hans Saner's *Kant's Political Thought* (Chicago: University of Chicago Press, 1973).

what Kant at one point calls mere "generality" (GR 42 [424]), not harmony. Only the free, universal acceptance of policies implicit in the legislative interpretation produces the harmony Kant has in mind.

Added to the political significance of these other writings on ethics, there is the explicitly legislative import of the third formulation of the imperative offered by Kant in the *Grundlegung*. This formulation reads: "Act according to the maxims of a universally legislating member of a merely potential realm of ends" (GR 57 [439]). The realm (or kingdom) of ends (*Reich der Zwecke*) involves for Kant the complex political idea of a merely hypothetical legislative order where every member is both sovereign and subject, maker and obeyer of common laws. It is not necessary to develop this idea fully to see its clear relationship to the legislative interpretation of the first formulation I am proposing.[7] If we recall that Kant firmly insists that the three ways he offers of presenting the imperative are "only so many formulas of the very same law" and that each one "unites the others in itself" (GR 54 [436]), we have solid grounds for believing that the first formulation is meant literally to involve a legislative test of maxims.

It may be objected that despite Kant's assertion about the three formulations saying the same thing, a serious problem is presented by the second formulation. This stipulates that rational nature is "an end in itself" and thus "must serve in every maxim as the condition restricting all merely relative and arbitrary ends"

[7] Thomas E. Hill discusses the possible moral significance of this formulation in his article, "The Kingdom of Ends," *Proceedings of the Third International Kant-Congress* (Dordrecht, Holland: D. Reidel, 1972): 310–315.But Hill does not clearly perceive that there may be a legislative dimension to the first formulation also. In a later discussion he explicitly states his belief that the two formulations differ in their practical application, with the first formulation involving a "stricter method" of reasoning. Thomas Hill, "Kantian Constructivism in Ethics," *Ethics* 99 (1989): 766.

(GR 54f. [436]; also 47 [429f.]). Since this seems clearly different from the other two formulations, is it not reasonable to suppose that each formulation contributes something new to the conception of imperative? If so, then despite what Kant says about the three formulas being identical, each successive formulation significantly amplifies the preceding ones. This reading of the *Foundations* supports the common interpretation that the first formulation involves nothing more than the formal universalizability of maxims, the second introduces values altogether lacking in the formal conception and somehow independently derived, and the third introduces a legislative test combining these two.

A proper reply to this familiar reading of the *Foundations* requires a full account of the mutual relationship of various aspects of a legislatively understood imperative, something I can only sketch here. The key point is that even conceived legislatively, the categorical imperative can be regarded in several ways. First, there is the formal aspect of the principle, in this case a rule requiring not just the universal *applicability* of maxims, but their universal *acceptability* as well.

Second, there is the material aspect of this principle—literally those substantive ends or values necessarily associated with a process of universal legislative choice. Beyond its formal procedural bylaws, any legislative process involves a determinate content: the interests and objectives of the legislators involved. What might these be in the case of a universal rational community? After more extensive argumentation I think it can be shown that since every rational person presumably protects her or his most vital interests in any free and unanimous voting procedure, these necessary ends will be those that are central to each person's rational agency and pursuit of possible purposes. As Alan Gewirth and others have argued, where human beings are concerned (and we should note that in presenting this formulation Kant often moves between the terms humanity and rational nature), these ends will almost certainly include the basic "generic goods" essential to human life (i.e., goods such as life, liberty of person and belief, freedom from injury and access to basic material necessities).[8] Each legislator thus makes "rational nature an end in itself" in protecting her or his own most basic ends as a rational (human) being, and, because of the constraints of the unanimous choice situation, in also necessarily valuing the fundamental interests and ends of every other rational (human) being. Understood this way, the formulation of "rational nature as an end in itself" is not the imprecise poetic expression it is sometimes taken to be, but is potentially a rigorous method of reasoning in which the fundamental and compatible aims of the community/ies of rational persons involved are identified, ranked and then utilized in a further process of legislative moral choice.

Finally, these two considerations—the formal and material aspects of the principle—come together in a unified third formulation which puts maxims to the test that they be capable of legislation by the community of rational persons, Kant's "realm of ends," where each person is both subject and object of the lawmaking process and where each proceeds to legislation by abstracting from his or her particular ends and taking into account everyone's most fundamental ends and purposes (the ends pertaining to "rational human nature" per se).

One of the recognized difficulties of the universalizability interpretation is that, apart from the vague common appeal to rational consistency, it produces no clear relationship between the first and subsequent formula formulations of the imperative. It also runs directly counter to Kant's strong assertion that the three formulations are different ways of saying the same thing. The fact

[8] Alan Gewirth, *Reason and Morality* (Chicago: University of Chicago Press, 1978) 52f. A similar list of basic ends is developed by Bernard Gert in connection with his conception of rationality. See his *The Moral Rules* (New York: Harper & Row, 1970), chapters two and three, and *Morality* (New York: Oxford University Press, 1988), chapters two and three.

that the legislative interpretation supports Kant's explicit statement regarding the relation between these formulations is, I think, important evidence in its favor.

The preceding remarks show that the first formulation of the categorical imperative can, with substantial fidelity to Kant's writings, be interpreted in the literally legislative sense requiring a proposal (maxim) to be capable of passage into law by and for the whole community of rational persons. In this second section, I want to depart from textual matters to argue independently that Kant's repeated claims that the categorical imperative is a necessary and sufficient guide to moral judgment become most compelling when the imperative is viewed this way. The best way of indicating the adequacy of the legislative interpretation is to contrast it, briefly, with the same imperative understood as a universalizability test. The difficulties with the latter are relatively obvious and have frequently been noted. They reduce concisely to the problem that the universalizability test sometimes produces morally bizarre conclusions because it renders moral principles a function of the particular volitions and circumstances of the choosing agent.

This problem appears in the case of actions whose universalized maxims do not involve a contradiction in conception and whose moral legitimacy, therefore, depends on the further test of the agent's coherent volition. Kant openly admitted that such actions are found in the case of imperfect or meritorious duties, such as the duties of mutual aid and self-development. Universalized maxims in this area, he tells us, are conceivable but the resulting behavior may nevertheless be immoral if it somehow involves a contradiction in willing. Logically universalizable but potentially immoral maxims also appear to crop up where some imprescriptible or perfect duties are concerned. This is especially true in the case of maxims based on what Christine Korsgaard calls natural as opposed to conventional actions.[9] A maxim permitting the killing of innocent persons is an example. Unlike false promising it is possible to conceive of this maxim universalized.[10] The question then becomes whether it is possible to will this maxim as universal law.

The problem with the universalizability interpretation is that it seems to allow some maxims to pass its test despite the fact that the resulting practices violate some of our strongest intuitive moral judgments.[11] One undramatic example concerns the kind of individual who, because of her nature or situation, feels no need of the assistance of others. Forced to choose between inconveniencing herself in a minor way on others' behalf (for example, by throwing a safety line to a drowning person) or proceeding on her way undisturbed, can this individual rationally choose to withhold aid, especially if she knows that because of the constraint of universalization she denies herself similar assistance in the future? Despite efforts to show that one could not rationally withhold aid in such cases,[12] it seems reasonably clear that anyone

---

[9] Christine Korsgaard, "Kant's Formula of Universal Law," *Pacific Philosophical Quarterly* 66 (1985): 25–47.

[10] In her *Acting on Principle,* 79f., Onora Nell argues that the contradiction in conception test rules out wanton killing because "[T]he normal and predictable results of generalised killing is not a means a rational agent could choose in order to act on the maxim of killing others." But Nell never clearly shows why it is irrational for some persons to opt for a world in which they preserve a power to kill.

[11] Hill offers an incisive review of the problems associated with the universalizability test in his article, "The Kingdom of Ends," 310f.

[12] In his *Generalization in Ethics* (New York: Russell & Russell, 1961) 269ff., Marcus Singer claims that, "no one, no matter how wealthy, strong or self-sufficient can so order and determine things as never to be in need of help of any kind or degree" (269). This may be true, but it does not follow that some rational persons might not choose to forego help of any kind if they could best serve their foreseeable interests by doing so. For a similar effort to affirm the irrationality of a rule denying mutual aid, see Barbara Herman, "Mutual Aid and Respect for Persons," *Ethics* 94 (1984): 577–602.

who greatly values her own freedom from interference, who lacks sympathy (perhaps because of a noble, stoical nature), or who, for various reasons, has no fear of falling victim to similar difficulties in the future, could rationally refuse help to others in cases like this. Just as I often decide to forego one good in the name of another, so an individual constrained by the requirement of consistency but knowledgeable of her circumstances might prefer convenience to a rule of mutual aid. In fact, on two occasions Kant mentions rational individuals entertaining just this kind of choice (GR 14, 48n.14 [398, 430n.]).

It might be objected that the individual who reasons this way is permitting too many subjective and particular considerations to influence her judgment. As a reminder to reflect on our true rational interest and not be swayed by the impulse of the moment, this is important. But clearly it is no objection to the form of reasoning presented here since the universalizability requirement in no way excludes the influence of personal information and values, but only requires us to accept *others'* acting on the maxims we propose. In brief, the universalizability test is entirely subjective and relative to the particular agent, and this is why it produces results contrary to our considered moral judgment.

We reach the same conclusion in the even more dramatic case of a maxim allowing the wanton killing of innocent persons. This maxim can logically be universalized without contradiction. On the universalizability interpretation, therefore, the question becomes whether a person can bring such a world into being without creating a contradiction in his or her willing. Despite strained efforts to suggest that one could not do this, there seems no reason why some rational persons could not will a world based on this maxim.[13] Particularly able gunslingers, for example, might

[13] Christine Korsgaard, in her article "Kant's Formula of Universal Law": 41ff., argues that the use of violent means for achieving one's ends involves a contradiction in willing because it leaves the agent insecure in the possession

find it rational to choose a world in which their increased risk of being killed is more than offset by their enhanced liberty of action.

How does the legislative test compare in this regard? The answer to this question requires some preparatory discussion since the test I am defending, though easily understood, has complexities we have not yet considered. We might begin by recalling that where concrete moral decisions are concerned, the test involves the following question: "Can I rationally believe that what I propose to do might be accepted as a valid law governing our behavior by all other rational beings?" In the case of our cold-hearted and self-sufficient individual, the question becomes: "Can I rationally believe that all rational individuals (including myself) would agree that it is permissible for people to avoid modest inconvenience by abandoning a drowning person?" In the case of killing the innocent, the question is "Can I rationally believe that all rational individuals (including myself) would agree that it is permissible for people to kill others whenever they please?"

One problem posed by both these questions is that it is not clear that "all rational persons" could agree to anything, much less to the acceptability of a specific maxim. Even the delimited group of rational persons known as human beings displays a bewildering variety of basic beliefs and values, and not even the presumably most objective facts of science (the earth's roundness, for example) are accepted by everyone. This legislative test of maxims, therefore, seems shipwrecked at the outset on a problem familiar to political life: the unanimity rule of social decision almost invariably leads to deadlock.

As serious as this problem may seem, it has been handled rather adroitly and in a roughly similar way by a number of

---

of the very goods he or she wishes to secure. Korsgaard seems to perceive the lack of conclusiveness in this argument when she acknowledges that the test of practical contradiction does not always have "complete success" in dealing with natural actions. Ibid., 43.

contemporary philosophers, including John Rawls, Bernard Gert, and Alan Gewirth.[14] Recognizing that the requirement of universal acceptability lies at the heart of any rational justification of principles, but also recognizing the virtual impossibility of unanimity under ordinary conditions, these writers have concluded that if there is to be rational justification at all, principles must be thought of as based on the judgments of persons who have been systematically deprived of all private knowledge of the particularities that separate them from one another. That is, principles must be accepted or "voted on" by radically impartial rational agents. Such agents' knowledge need not exclude generally known particular information (for example, the fact that individuals hold certain special beliefs or values), but it must not include information that would allow any particular rational agent to know which beliefs or values are his or hers. Understood this way, moral choice and the "legislative" test I have been outlining become equivalent to choice by a community of radically impartial persons, or, what is the same thing—since all these voters face the same complex decision—choice by a single person who assesses the array of existing differences and particularities from an impartial perspective.

It might be objected that this step in reasoning effectively eliminates any truly "legislative" dimension to the test I am proposing, rendering it more like a procedure of impartial rational choice than any kind of full-bodied legislative process.[15] This objection is valid to the extent that this interpretation involves no real political bargaining or negotiation (although it does require a

---

[14] John Rawls, *A Theory of Justice* (Cambridge, Massachusetts: Harvard University Press, 1971); Gert, *The Moral Rules* and *Morality;* Gewirth, *Reason and Morality.*

[15] A similar objection against Rawls's use of the contract metaphor for what is, essentially, a process of individual rational decision is voiced by Jean Hampton. See her "Contracts and Choices: Does Rawls Have a Social Contract Theory?" *The Journal of Philosophy* 77 (1980): 315–88.

single impartial individual to engage in a delicate process of weighing goods and evils as they affect different persons). Nevertheless, the term "legislative" is worth retaining here because this procedure begins with the requirement of universal acceptability. The fact that this requirement leads one to invoke a procedure of impartial decision (as a way of achieving unanimity) is a matter whose necessity has only been made clear by contemporary ethical discussions. Indeed, one reason why the legislative nature of the first formulation has not previously been perceived may be because, until recently, we lacked the conceptual apparatus (better understanding of the procedure of impartiality) needed to evidence the workability of the formulation understood this way.

We might note that this legislative reading of the first formulation and its reduction to a principle of impartial rational choice has strong affinities with a number of Kant's other key ethical ideas. The most important of these is his conception that the categorical imperative (as opposed to hypothetical imperatives) excludes all influence of particular or subjective inclinations. It would be foolish to maintain that Kant excludes a role for human ends and desires in the application of the categorical imperative—in fact he repeatedly alludes to specific human needs. We may presume, therefore, that it is the decisive influence of *particular* desires and inclinations that he means to bar from the moral reasoning process. A very general interpretation of all the formulations of the categorical imperative in terms of impartiality has been suggested by Rawls—but not in terms of the prior idea of universal *legislatability*—and a specific idea of legislative impartiality has been ascribed to the "kingdom of ends" formulation by Thomas Hill.[16] We are not unwarranted,

---

[16] Rawls, *A Theory of Justice,* Section 40, 251–57 and his Dewey Lectures, "Kantian Constructivism in Moral Theory," *The Journal of Philosophy* 77 (1980): 515–72; Hill, "The Kingdom of Ends": 307–15.

therefore, in assuming that for Kant himself the legislative process intended by the first formulation also implies a concept of impartial rational choice.

It might be objected here that if the difficulty of "deadlock" in the legislative interpretation can be solved by the imposition of impartiality, so can the difficulties in the universalizability view. Why not argue, then, that Kant interpreted the first formulation in terms of universalizability with an unstated assumption of reasoning from an impartial standpoint? Although I cannot rule out this approach, several considerations militate against it. Foremost is the fact that there is no natural connection between universalization and the idea of impartiality. Why then should we suppose in the absence of any explicit indication that Kant meant himself to be understood this way? In contrast, the legislative interpretation leads quite naturally to the idea of impartiality. After all, real legislative processes, with their long-term focus and pressure toward agreement, almost always involve some measure of impartiality and a willingness to "take other persons' interests into account." Together with the strong legislative overtones of the first formulation, this suggests that Kant arrived at the assumption of impartiality via the legislative approach rather than through the idea of universalizability.

If all this is correct, we are now able to give our legislatively construed principle final formulation. This is: "Act only on those maxims which you can rationally propose would be accepted by rational persons deprived of particular knowledge and rendered impartial before the array of existing facts and values." Applied to our earlier question about mutual aid, the question now becomes: "Can I rationally propose that rational persons would accept (vote for) a law permitting the denial of aid in cases like this?" Since we know nothing about potentially vast segments of the rational community, we are certainly permitted to limit our thinking here (as does Kant) to human beings, both because this is the only rational community with which we are familiar, and because we can suppose that other rational beings would permit us our special law in this matter so long as it did not contravene more universal legislation we might someday promulgate. Thus, our question boils down to the following: "Is it rational to believe that impartial and rational human beings would consent to this denial of mutual aid?"

Put this way, I think the answer to this question is evidently negative. The overwhelming majority of human beings, after all, greatly value assistance in cases like this and to secure it are more than willing to tolerate the slight inconvenience involved. The beliefs and values of a few cold-blooded, self-sufficient persons are not irrelevant to such an impartial reasoning process. They are considered and assessed impartially, and they help furnish one argument against a rule of mutual aid. Nevertheless, from an impartial standpoint it is certainly rational for us to try to protect those very basic values essential to our welfare and pursuit of possible purposes. Life is among these and, regarded impartially, its preservation has far higher priority than convenience. Consequently, I think it is rational to believe that impartial rational persons would require giving assistance in cases like this. The same reasoning applies to the even more extreme rule permitting the killing of innocent persons. However rational a particularly able killer might find such a rule, our overwhelming valuation of life indicates that this is not a choiceworthy rule for one deprived of knowledge of his or her particular circumstances. The legislative interpretation of the categorical imperative thus supports two common sense moral rules, and conforms to Kant's own conclusions on these matters.

It may be objected here that this way of reasoning seems to make basic moral rules depend on the empirical facts of human valuation and sentiment in ways that Kant, with his insistence on morality as involving universality and necessity, would not permit. But we must be very careful. In fact, the fundamental formal rule being applied in this case—the rule of universal acceptability as law—has just the necessity Kant claims for it. As the supreme moral principle it defines the unique condition of a free harmony of wills, and, as such, the unique condition of morality. As far as the derivative rules of mutual aid or non-killing are concerned, they have derivative necessity given the facts of human nature. I will not speculate on whether such derivative moral laws would be abandoned were human nature to change in some fundamental way, but I would point out that Kant does not appear to claim that these specific items of moral legislation necessarily extend beyond the community of rational *human* beings. In fact, apart from the categorical imperative itself, he seems to ascribe universality only to those derivative moral rules, such as promise keeping, whose willed infraction involves conceptual self-contradiction.

I might also add that these considerations throw open to question the common classification of Kant as a deontological as opposed to a contractualist or teleological thinker. In fact, elements of all three approaches have their place within a legislatively construed categorical imperative. The contractualist dimensions are evident, as is a teleological concern with the impact of various legislative proposals on the welfare of persons, with their happiness. Since the resulting laws override particular considerations of personal or group happiness, however, this imperative has a distinctly deontological quality. Indeed, because the implicit legislative community comprises all *rational* beings whose thinking is not determined by personal interests, its rules are properly regarded as objective laws binding on all rational persons.

We are now better able to see why this legislative test works where the universalizability test does not. Only this interpretation leads directly to and embodies the kind of strict impartiality that is acknowledged to be a key feature of moral reasoning. Oddly enough, the universalizability requirement is often carelessly described as "the principle of impartiality," perhaps because it prohibits individuals from making unreasoned exceptions in their own cases.[17] But clearly this is a very limited kind of impartiality since the principle in no way rules out reasoning from highly idiosyncratic and particular perspectives. In contrast, the legislative test excludes all particular inclinations (unless they are impartially assessed) and thus most closely approximates the "moral point of view." This explains its conformity to our strong intuitive judgment in this case, and, if the applied work of thinkers such as Rawls, Gert, and Gewirth is to be relied on, in many other cases as well.[18]

---

[17] D.H. Monro points out in his article, "Impartiality and Consistency," *Philosophy* 86 (1961): 176, that the rational requirement of consistency, which merely prohibits exceptions not allowed to others in like circumstances, is often confused with the moral principle of impartiality, which rules out most exceptions in one's own favor.

[18] It lies beyond the scope of this essay to demonstrate the value of this method of impartial rational choice in deriving a complete set of interpersonal or institutional moral principles and ranking them in cases of conflict. Nevertheless, I believe the work of Rawls, Gert, and Gewirth indicates the general fruitfulness of this method. We might anticipate, of course, that in certain cases, where not all facts are at hand or where incomplete impartiality with regard to any matter in dispute cannot be imposed, no single moral conclusion will result.

But it is one thing to acknowledge that not all disputes will be immediately resolvable and quite another to conclude that a legislative-impartiality approach is unable to arbitrate moral conflict. Remarkably, Keith Ward uses this argument in his *Development of Kant's View of Ethics* 113–18, to dismiss as completely formal and "unable to provide a sufficient criterion for morally right action" a legislative formulation of the categorical imperative. Viewing this formulation as involving "antecedent colegislation" Ward comes closer than most writers to seeing its legislative significance. But because he believes it will leave many questions unanswered, he rejects it as a sufficient criterion of morality and moves on to other formulations in search of more determinate content for the moral reasoning process.

There are good reasons, then, for believing that, understood this way, the categorical imperative displays the very qualities Kant repeatedly claims for it. On the one hand, precisely because it requires nothing more than that conduct judged to be moral be acceptable to all rational persons, it is the supreme principle necessarily accepted by all rational persons. In fact the principle merely spells out and renders as a test what is involved in the idea of morality and rational justification in the first place. On the other hand, the principle provides a single, high-order rule invariably capable of distinguishing moral from immoral maxims. It is, in short, the necessary *and* sufficient guide to conduct which the universalizability requirement is not, but which Kant claimed the categorical imperative to be.

The fact that this legislative interpretation can be shown to display qualities Kant attributed to the categorical imperative suggests that this is the interpretation he had in mind. Added to the previous textual considerations, I believe this makes it reasonable to suppose that, however unclear Kant may sometimes have been, he understood the first formulation to involve a legislative test whereby maxims are judged in terms of their *acceptability* as law to each and every member of the community of impartial rational persons.

2

Kant on Christian Love

If one were seeking insight into the nature and meaning of Christian love, Immanuel Kant would hardly seem to be the first philosopher one would consult. Kant, the rationalist foe of all religious enthusiasm and emotion in the moral life, the advocate of unflinching devotion to duty, does not strike one as a thinker likely to understand the meaning of religiously inspired, self-emptying love. Indeed, there is some basis for the intuition that Kant's moral thinking may not be the best introduction to Christian ethics, and that the categorical imperative and the Christian idea of *agape* may be fundamentally different guides to moral conduct. Nevertheless, Kant did write about Christian love. His reflections on its meaning remain of interest, and those seeking to understand the nature and limits of *agape* can still perhaps learn from Kant.

In what follows, I want to explore Kant's specific efforts to interpret the love commandment found in Matthew 22:37–40 and in Mark 12:30–31. These efforts, often brief and always undertaken in the context of other discussions, form the principal resource for anyone wishing to comprehend Kant's understanding of *agape*.[1] In addition, Kant's writings contain several extended discussions of benevolence and beneficence, duties that he sometimes equates with Christian neighbor-love. For this reason, I propose to

---

[1] Although Kant refers on several occasions to the love commandment found in Matthew 22:37–40, he never actually uses the term *agape* in discussing Christian love.

examine these discussions as well. But before turning to either matter, it is important first to dispel a popular misconception of Kant's ethics that renders his thinking not merely distant from the idea of *agape* but absolutely hostile to any ethic involving human sympathy and fellow-feeling.

The misconception derives from Kant's well-known discussion of the (morally good) will at the beginning of his *Foundations of the Metaphysics of Morals.* In the effort to develop the idea of action done *from* duty, Kant dismisses as having "no moral worth" a kindly deed performed by a person who finds "inner satisfaction in spreading joy" to others. In contrast, Kant holds up the example of an individual with little sympathy in his heart, "by temperament cold and indifferent to the suffering of others," who nevertheless is able out of duty to benefit others in distress. Such a person, Kant maintains, finds in himself a far higher source of moral worth than can be gained through a good-natured temperament (GR 15 [398]).

This passage has disturbed or angered generations of readers. The poet Friedrich Schiller used it to characterize Kant as a philosopher who regarded the malign performance of outwardly moral acts as superior to generous compassion for one's fellow human beings.[2] If Schiller and others who have read the passage this way were right, Kant's ethics would be infinitely remote from anything like the kind of other-regarding compassion that many believe to be an important part of Christian love.

But Schiller and others miss Kant's point here. Kant's aim, after all, is to illustrate what it means to act *from* duty, to be essentially motivated by moral concerns, and he wishes to distinguish such action from that which happens to conform to duty but is without genuine moral motivation, what he calls

[2] *Über Anmut und Werde.* For a fuller discussion of the relations between Schiller and Kant see Lewis White Beck, A *Commentary on Kant's Critique of Practical Reason* (Chicago: University of Chicago Press, 1960) 120n. 231.

action *in accord with* duty. To this end, Kant performs a thought experiment in which he offers what today might be called a "controlled sample" of genuine moral motivation.[3] For this purpose, actions motivated by feelings of sympathy and benevolence to others will not do. They represent "tainted" specimens since it is difficult in such cases to determine whether the action is motivated by sentiment and feelings or by a genuine respect for duty. Kant finds his 'pure' instance of moral conduct, therefore, in the case of an individual respectful of the moral law and of whom one can predicate no emotional or self-oriented reasons for conduct. Here there is no doubt, Kant concludes, that respect for duty is present.

What Kant clearly is *not* doing in this passage is holding up as the paragon of moral virtue someone who dislikes or hates those he helps. Nor is he in any way belittling the conduct of individuals whose sense of duty is supplemented and enhanced by feelings of sympathy and love. Indeed, we shall later see that Kant believes it is our duty to encourage and to develop such benevolent feelings. Instead, what Kant is trying to do is clarify what he believes to be the essential basis or determining ground (*Bestimmungsgrund*) of moral worth. For reasons we shall soon explore, he insists that this cannot be found in emotional states or preferences, but must be rooted in a reasoned commitment to duty. The purpose of his thought experiment, therefore, is to separate a genuine moral motivation from adventitious motives that can obscure its presence or its absence. How unfortunate that for many readers, this "pure" case should be taken as Kant's moral ideal.

Kant's point in this passage will not, of course, entirely make sense to those who believe that various emotional states—

---

[3] In KPV (164) Kant refers to this moral stance as one involving "negative perfection" and he explicitly advocates its use in illustration as a method of moral pedagogy.

sympathy, compassion, or love—are properly the basis of moral conduct and moral worth. Why Kant dismisses these states as bases for morality, as appropriate determining grounds for the will, requires further explanation. As a topic, it also serves as introduction to Kant's own conception of Christian love. A brief passage in the *Foundations* is crucial for our understanding of Kant's perspective. Continuing his effort here to distinguish between illegitimate motives based on feelings and the only valid motive based on duty, Kant reasserts his conviction that only an action done out of respect for the moral law has moral worth. He comments:

> It is in this way, undoubtedly, that we should understand those passages of Scripture which command us to love our neighbor and even our enemy, for love as an inclination cannot be commanded. But beneficence from duty, when no inclination impels it and even when it is opposed by a natural and unconquerable aversion, is practical love, not pathological love; it resides in the will and not in the propensities of feeling, in principles of action and not in tender sympathy; and it alone can be commanded. (GR 15f. [399])

It alone can be commanded. These words express the heart of Kant's objection to any ethic based on emotional preferences and to any understanding of Christian love that would interpret it in terms of emotional states. For Kant, a defining feature of morality is its imperatival force: we encounter moral rules or principles as commands that we are called on to obey. Moreover, the whole language of moral praise and blame expresses this same understanding. We commonly exhort one another to moral conduct, and we condemn and discourage immoral behavior. But what is commanded must be capable of being done. "Ought implies can" because it makes no sense to demand what is beyond a person's power. For Kant, it follows that duty cannot depend on or proceed from emotional states—however refined or lofty these might be—because he believes that emotions are essentially beyond our control. I feel sympathy for another or I don't; I love or don't love. As emotions, these are states I experience, suffer or enjoy, but I cannot simply elicit them by acts of will.

For this reason, Kant believes that the command to love the neighbor, understood as a command to express feelings of compassion or concern, would be nonsensical—like a command to enjoy food or music that one finds distasteful. In Kant's view, an ethic based on feelings would not only run counter to our shared conviction that moral duties are something we can be called on to obey, but it would also end by violating some of our deepest beliefs about moral praise or blame. Such an ethic would lead us to praise persons for harboring feelings as natural to them as their tastes and for which they are not really responsible, while we might be led to condemn individuals of a different make-up unlucky enough not to enjoy or experience these feelings. For Kant, these absurdities lead to the conclusion that where love is regarded as a matter of feeling, "a *duty to love* is an absurdity" (MS 203)

Nevertheless, the fact that the injunction to love the neighbor cannot be understood as a requirement of emotional concern for that neighbor does not mean that the Gospel requirement lacks sense in Kant's eyes. In place of such a "pathological" or feeling-based idea of love, Kant interprets the Gospel command as involving "practical love." This resides in the will, not the feelings, and calls forth action instead of sympathy. In the *Foundations* Kant states matters this way, but he does not really develop the idea. Not until the *Critique of Practical Reason* do we encounter a fuller discussion of the commandment to love God above all and the neighbor as oneself (KPV 85f.). Commenting on this, Kant begins by insisting that love of God as an "inclination" or state of emotional desire is impossible since God cannot be an object of the senses. Presumed here is Kant's belief that all such desiring love has a sensual or sensory basis. But since God cannot possibly be loved in this way, to love God can only mean that we endeavor to fulfill all His commandments (which Kant everywhere equates with obedience to the dictates of the moral law) (Rel 148). While it is possible to love one's fellow human being in a "pathological" or feeling-determined way, such feelings cannot be commanded. This means that love involves the requirement that we try to practice all our moral duties with respect to the neighbor. In sum, Christian love for, Kant represents an injunction to the most perfect fulfillment of rationally understood moral duty. It involves respecting the dictates of the categorical imperative in their fullest and most complete sense.

Anyone nourished by the tradition of modern Christian reflection on *agape* might well find this an impoverished account of the love commandment. Altogether lacking in Kant's thinking is any sense that love might involve the kind of unnecessitated and overflowing regard for the other that is emphasized by a writer like Anders Nygren. Also missing are the themes of self-sacrificial devotion stressed by writers like Kierkegaard or Reinhold Niebuhr. Nor does Kant appear to discern any complexity in the relationship between the twin commandments of love of God and love of neighbor, since he collapses both commandments into the same requirement of respect for rational duty. Within this scheme, for example, the vertical element of relationship with God does not significantly inform the horizontal realm of the ethical. In the manner lamented by Kierkegaard, God becomes an "invisible vanishing point" (FT 68).

These are possibly serious limitations in Kant's view. But it would be wrong to characterize his interpretation of the love commandment as either ethically or religiously impoverished. On both the moral and religious planes, Kant's perspective has a dimensionality and depth that merits attention. Morally, for example, there is abiding validity in his critique of any interpretation of Christian love that renders it a matter of sentiment rather than a practical commitment to forms of conduct. On the moral level, as well, Kant's equation of Christian love with duty in its fullest sense also represents a response to the many problems that accompany alternative interpretations. Gene Outka has identified a number of these problems.[4] How do love and justice relate to one another? When do the needs of the person who stands before me take precedence over those of the community as a whole? What degree of self-regard is permitted to one committed to altruistic concern for others? Is it possible for totally selfless devotion to others to violate valid requirements of self-respect and specific duties to oneself? While the equation of Christian love with full obedience to the moral law may not immediately answer all of these questions, it does point to their solution. Viewed this way, the agapistic response would not be one-sidedly altruistic, individualistic, or self-sacrificial, but would rather conform to that fair and reasonable reconciliation of competing interests at which morality aims.

Religiously, as well, Kant's discussion of love merits attention, because it forms the basis for an important effort on his part to provide a philosophical understanding of the problem of sin. Kant's development of this theme does not occur until his *Religion within the Limits of Reason Alone.* There, to the surprise of many readers who regard him as an optimistic enlightenment philosopher, Kant introduces the idea of "radical evil" in human nature, and he goes as far as to suggest a role for divine grace in the economy of human moral redemption.[5] This mature treatment

---

[4] Gene Outka, *Agape: An Ethical Analysis* (New Haven: Yale University Press, 1972).

[5] Reinhold Niebuhr, *The Nature and Destiny of Man,* 2 vols. (New York: Charles Scribner's Sons, 1941) 1; 120n. 12. In *The Nature and Destiny of Man,* Reinhold Niebuhr takes note of Kant's discussion of "radical evil" but dismisses this as "contradictory" to Kant's whole scheme of thought and as a pietistic intrusion in the thought of an Enlightenment optimist.

of sin is given its earliest expression by Kant in the *Critique of Practical Reason,* where it is intimately connected with Kant's treatment of Christian love.

Kant's discussion here anticipates in many ways Niebuhr's understanding of love as an "impossible possibility." On the one hand, says Kant, because of the unyielding stringency of its demand for full moral obedience, the love commandment provides a stimulating challenge to the moral life. The Matthean commandment, he states, presents an "ideal of holiness" so perfect that it is "unattainable by any creature." Nevertheless, in this very purity, the commandment is an "archetype which we should strive to approach and to imitate in uninterrupted infinite progress" (KPV 86).

On the other hand, this lofty ideal not only stimulates and challenges, it also places human striving under judgment and brings a proper sense of humility to the committed moral agent. In this connection, Kant criticizes the fancied moral perfection and "self-conceit" of those who believe love to involve a spontaneous goodness of heart. Such persons, Kant says, think that they can fulfill the moral law by intensity of feeling and by occasional acts of self-sacrifice. But duty has nothing to do with feelings, and its dictates are not fulfilled by fleeting instances of generosity. Instead, duty places a stern and unremitting discipline on the whole of life (LE 135–37). This discipline runs counter to many of our ordinary desires and to our natural forms of self-love. The immediate effect of the love commandment, therefore, is to humble and chasten us. More profoundly, it drives us to repentance for our repeated moral failures and to a sense of dependence on God's mercy (combined with our own striving) for the perfection of our moral natures.

In a long footnote near the end of the second *Critique,* Kant highlights this last theme by contrasting Greek ethics, marked as it is by a confidence in human beings' this-worldly attainment of the good, with Christian ethics and its reliance on a transcendent religious fulfillment of human's destiny. The difference between these ethics, he remarks, follows strictly from the stringency of Christianity's commitment to love:

> Christian ethics, because it formulated its precept as pure and uncompromising (as befits a moral precept), destroyed man's confidence of being wholly adequate to it, at least in this life; but it reestablished it by enabling us to hope that, if we act as well as lies within our power, what is not in our power will come to our aid from another source, whether we know in what way or not. (KPV 132, n. 2)

The concluding note of hesitation here about the extent to which human beings can understand God's morally redemptive grace is typical of all Kant's thinking on this matter. Even in the *Religion,* where the second *Critique's* idea of reliance on an infinitely prolonged afterlife as a means of attaining perfect virtue seems to be replaced by the idea of a more direct and immediate divine act that heals the will, Kant always appears uneasy with the direction in which his rigorous analysis is taking him. As commentators have noted,[6] his discussion there wavers between a stress on man's clear and unaided duty to do all that is in his power on behalf of the moral law and a qualified reliance on God's aid. The details of Kant's agonizing engagement with the problem of divine grace need not concern us here. What is important is that a thinker, who many have characterized as hostile to Christian theology, is, in some ways, one of the great philosophical interpreters of the Christian doctrine of sin. What is also especially noteworthy in this context is that his development of these ideas builds directly on his understanding of Christian love.

---

[6] See, for example, the discussion of Kant's doctrine of grace in Allen W. Wood, *Kant's Moral Religion* (Ithaca, New York: Cornell University Press, 1970) 208–248.

So far, I suggest that Kant equates Christian love with strict respect for duty. This presumably means firm resolve never to do what duty prohibits as well as a commitment to some positive duties. In addition to this general position, however, there are passages in Kant's writings where, in a more focused way, he equates love of neighbor with a very special positive duty that he calls "benevolence" or "beneficence." Kant's terminology here is not always clear. On some occasions he speaks of the duty he has in mind as benevolence, although, in clear discomfort with this word's dispositional connotations, he sometimes calls it "benevolence, as conduct" (MS 203). Elsewhere, he resorts to the more action-oriented term "beneficence" (MS 196).

In all these references, what Kant has in mind is not merely sympathetic concern for others—since that involves uncommandable feelings—but an active, willed effort on our part to aid our fellows. In Kant's view, beneficence involves our making others' happiness our end. It finds concretion in the duty "to promote, according to one's means, the happiness of others in need, without hope for something in return" (MS 247; also 303). So understood, this is a positive duty without fixed limits. Along with other acts of indefinite obligation, like the requirement of self-development, it belongs to the domain of "meritorious" duty. In this respect it differs from an "imprescriptible" duty like promise keeping, whose violation is always clear and specifiable (GR 42 [424]). But the fact that we cannot in a general way identify violations of this duty does not make it purely supererogatory. Within the limits of our means and abilities, we are each morally required to benefit and aid our needy neighbor. We may not be required to impoverish or kill ourselves in this effort (MS 248), but we are called upon as a matter of moral obligation to check our pursuit of self-interest when others' welfare can reasonably be served.

What does Kant believe to be the moral and conceptual basis of this duty? How is it justified? Expectedly, Kant makes no reference to specifically religious justifications in answering these questions. "Love of neighbor" as beneficence is not, for him, a consequence of membership in a self-designated community of saints. Instead, beneficence is an aspect of rationally defensible moral obligation and, as such, is incumbent on all who are human. Interestingly, however, Kant's specific efforts to develop the rational basis of this duty form one of the most problematical aspects of his moral theory. A brief look at these efforts helps shed light on the strengths or weaknesses of Kant's moral theory as a whole.

It is well known that for Kant, all conduct must conform to the dictates of the categorical imperative. This, in turn, involves the requirement that all proposed policies of action (or what Kant calls "maxims") be capable of becoming universal law. (Kant states this imperative formally as follows: "Act only according to that maxim by which you can at the same time will that it become universal law."[7]) However, this is largely a negative test: we may do whatever we please as long as it is not incapable of becoming universal law. To establish whether there is a duty of beneficence, therefore, one need only put the contrary maxim to the test. Is a policy of non-beneficence (of resolutely denying others needed aid when it is in our power to give it) capable of becoming universal law? If the answer to this is "no," then beneficence is in fact a basic moral obligation.[8]

---

[7] The translation here is my own. I believe both Beck and Paton mistranslate this formulation by interpreting the German *werde* as *should become,* a reading that is a valid English subjunctive but that also has the sense of making universal law dependent upon the willing agent's wishes, in keeping with a generalization view of the formula.

[8] For a discussion of the logic of moral permissions and requirements in a Kantian framework, see Onora Nell, *Acting on Principle* (New York: Columbia University Press, 1975) 77.

Obviously the answer to this question will depend upon what is meant by "universal law" and what further test is involved in determining whether a maxim is capable or incapable of becoming such a law. Commonly, both these matters are interpreted in terms of the test of "universalizability."[9] This involves two things: First, there is the requirement that I think of what I propose to do as being known by and open to all other, similarly situated moral agents. This implies that the action involved is of the sort that can logically be conceived as acted on by everyone. Second, once this is done, I must consider whether I would personally be willing to put up with others' acting in this way with respect to me. Since a maxim permitting nonbeneficence can logically be thought of as acted on by all persons, the categorical imperative seems to yield the following concrete question as a final test for this maxim: Would I be prepared to live in a world in which persons were morally permitted casually to deny others (including myself) needed aid when it is otherwise in their power to give it? If the answer to this is "no," then beneficence becomes a moral obligation.

This, in fact, seems to be the shape of Kant's own efforts to justify the duty of beneficence. In the *Doctrine of Virtue,* for example, he observes that, "every man who finds himself in need wishes to be helped by other men." But if anyone denies the obligation to beneficence, says Kant,

> ... if he lets his maxim of being unwilling to assist others in turn when they are in need become public, that is, makes this a universal permissive law, then everyone would likewise deny him assistance when he himself is in need, or would at least be authorized to deny it.

---

[9] See chapter one, note 1.

Hence, Kant concludes, "the maxim of self-interest would conflict with itself if it were made a universal law, that is, it is contradictory to *duty*." (MS 247)

In the *Foundations* a very similar argument is set forth. There we read that since instances can arise in which one needs "the love and sympathy of others," a person who denied the obligation to mutual aid would effectively rob himself of desired assistance and his will would be "in conflict with itself" (GR 41 [424]).

On the surface, these are convincing arguments. But, as many critics have noted, they do not withstand close scrutiny. Granted that if I choose to reject a duty of beneficence, in keeping with the requirement of universalization, I thereby deny myself the assistance of others in moments of need. But why can't this sometimes be a reasonable choice? After all, life is a matter of choice among competing objectives. We cannot have everything we wish and we must often trade off one valued good for another. If I choose to become a professor of ethics, for example, I probably cannot at the same time be an astronaut. Why, then, can't I, as a cold and self-sufficient person, choose to forgo aid from others in order to free myself from a similar responsibility to them and thereby to better pursue my other goals?[10]

Remarkably, Kant seems to recognize the limitations of the universalizability view. In a footnote in the *Foundations*, he brings up a negative version of the Golden Rule and stresses its inadequacy as a guide to conduct:

> Let it not be thought that the banal *"quod tibi non fieri, etc."* [What you would not wish done to you, etc.], could here serve as guide or principle, for it is only derived from the principle and is restricted to various limitations. It cannot be a universal law, because it contains the ground neither of duties to oneself nor of the benevolent duties to others (for many a man would gladly consent that others should not benefit him, provided only that he

[10] See chapter one, note 12.

> might be excused from showing benevolence to them). Nor does it contain the ground of obligatory duties to another, for the criminal would argue on this ground against the judge who sentences him. And so on. (GR 48 n. 14 [430 n.])

Kant's efforts rationally to justify this duty, therefore, seem to fail. What are we to conclude from this? There are several possibilities. One is that Kant is simply mistaken: beneficence is not a universal moral obligation. This might suggest that beneficence properly finds grounding only in the select domain of a religious ethic, where individuals evidence their gratitude for God's generosity by rendering assistance to the neighbor. For Christians, the failure of Kant's argument would thus serve as point of departure for a demonstration of the special link between Christian faith and neighbor-love.

But there may be another way of interpreting Kant on beneficence. The first formulation could involve more than a rule of universalization and a test of the resulting universalized maxim in terms of the agent's preferences. What if the second part of this procedure is not meant to be rooted in the conditions and needs of the specific willing agent? What if, in considering a maxim, I as the agent must also take into account the impact of that maxim on *other* persons with possibly different needs or in different circumstances? In other words, what if I had to assess my universalized maxim in an *impartial* way, as though I could be any of the persons (including myself) who might be affected by it? In that case, the results of the reasoning process might be far different from that obtaining when a purely subjective test is imposed.

We have already seen that it is entirely possible for me as a self-sufficient, cold-tempered individual who clearly knows my own interests to dispense with a rule requiring mutual aid. But is this possible for me as a person who must reason impartially and take into account—literally step into the shoes of—every other member of society? Most human beings, after all, occasionally need assistance from others and are more than willing to put up with the slight inconvenience of providing it in order to be able to rely on aid when we need it. Not knowing which of these many possible persons I am (in the fashion of the hypothetical contractors of John Rawls's "original position"[11]), is it not rational for me to select a rule of beneficence, especially if this is construed so as not to require me normally to undergo great risk? In this respect, the duty of beneficence is somewhat like the purchase of insurance, whereby one hopes that repeated small disbursements may protect one against the harm of catastrophic loss. Although some particular persons who are able accurately to predict their future experiences may find it rational to decline insurance or to self-insure, anyone rendered less knowledgeable (or "impartial") before the future will find insurance a rational purchase.

On this interpretation, then, we have a reading of the first formulation that, just as Kant says, yields the duty of beneficence. Furthermore, the suggestion that reasoning about morality takes place under conditions of impartiality has deep roots in Kant's ethics, especially in connection with his important distinction between the categorical imperative and hypothetical imperatives (GR 311f. [414f.]). The former is distinguished by the fact that it does not permit merely particular desires—as opposed to generalized human needs—to be made determining considerations of the will. A number of writers—including Rawls, Thomas Hill and Jeffrie G. Murphy[12]—have regarded impartiality as a key feature

---

[11] John Rawls, *A Theory of Justice* (Cambridge, Massachusetts: Harvard University Press, 1971) 118–192.

[12] Rawls, *A Theory of Justice,* Section 40; John Rawls, "Kantian Constructivism in Moral Theory," *Journal of Philosophy* 77, no. 8 (1980): 549–50. Thomas Hill, "The Kingdom of Ends," *Proceedings of the Third International Kant Congress* (Dordrecht, Holland: D. Reidel, 1972) 310f.; Jeffrie G. Murphy, *Kant: The Philosophy of Right* (London: Macmillan, 1970) 94, stresses the way in which Kant's view rests on rational choice utilizing the "essential ends" or

of Kant's ethical theory. Applied to the first formulation and to its justification of the duty of beneficence, this interpretation makes good sense.

A further way of reading the first formulation gives impartiality an important role in its application and is even more directly suggested by the wording of the formulation itself. This involves taking the phrase "universal law" in its fullest and most explicitly *legislative* sense. Commonly, this phrase is taken to mean that a proposed maxim be universalized as *binding* on all persons—a law for them. But we might here recall Lincoln's observation that a republic is not just *for* the people but *of* them and *by* them as well. This suggests that a universal law may not just be applicable to and binding on all rational persons, but may also be a law that has been freely accepted by, even voted for, by all such persons. It is *their* law in the sense that they have helped will it into existence—or could be expected to do so. With this explicitly legislative metaphor in mind, therefore, we may think of universalized maxims as having to pass a test in which they must receive not just my approval but that of all other rational persons convened in a hypothetical democratic assembly. Elsewhere, I have developed this conception at greater length and connected it with Kant's understanding of all three formulations of the categorical imperative.[13]

A legislative interpretation has the effect of highlighting the presence of the concept of impartiality we have just explored. To reason impartially, after all, entails placing all persons' individual interests (including one's own) in the context of an array of competing interests and then, on this basis, to select a course of action most likely to be adopted by others who possess the ability to vote against whatever endangers their most vital interests. But

---

common basic interests of humankind, an interpretation that relies on an impartial assessment of human values.

[13] See chapter one.

this is precisely what the legislative interpretation involves. As a moral agent reasoning this way, I must mentally "poll" the variety of interests in the hypothetical legislature as I reason toward what would be the likely outcome of a voting procedure in this context. I must effectively think like an impartial—or, better, omnipartial—spectator and ask which forms of conduct would reasonably be accepted as abiding rules of conduct ("laws") in the give and take of a legislative assembly where each person has a veto that he or she must use judiciously (since others' also possess a veto). Thus, by taking Kant's phrase "universal law" in its most naturally legislative sense, we are led back to the impartiality procedure which, as we have seen, is perhaps the best way of making sense of the first formulation.

What does this tell us about the duty of beneficence? Nothing more, perhaps, than what was already said in connection with the view that sees the first formulation as involving universalization plus impartial assessment of the resulting conduct. Both of these interpretations yield the same decision process, although the legislative interpretation, I would argue, does so more naturally and more in keeping with Kant's explicit remarks. In either case, however, beneficence emerges as an important human obligation, just as Kant says it does. Reasoning impartially and taking into account all possible interests, any of us would be foolish to dispense with a duty of mutual aid. Similarly, we must expect that a community of hypothetical legislators would readily "pass" a moral law establishing this duty since the vast majority of human beings are benefited by it and even a minority that might be inconvenienced is not seriously disadvantaged by its establishment. On either account, therefore, Kant is right. There is such a duty incumbent on us all. Beneficence and mutual aid need not be regarded as the moral property of any special religious group. And to the degree that active regard for the neighbor is morally a part of *agape,* we can say that Christian ethics here shares a commitment with an ethics based on reason and open to all who are human.

Much of what I have said to this point should be somewhat chastening to those who hold a high-flown conception of Christian love. Kant shows us, I think, that interpretations of Christian love that stress its affective dimension can both violate our understanding of moral obligation and can trivialize the moral and religious depths of this requirement. He also shows us that active commitment to others' welfare may not be a uniquely Christian requirement. If not construed to require completely sacrificial devotion, it can be rationally justified as an obligation for all who are human. He further suggests the difficulties of the Golden Rule as a moral principle, and, perhaps more importantly, he shows that full regard for all our neighbors' welfare requires a structured way of thinking in which valid but competing claims can be adjudicated and assessed.

If this was all that Kant had to say about neighbor-love, I believe his contribution would be substantial. Before concluding, however, I might mention one further idea of Kant's that deepens his understanding of the extent of our duties to others. I began this essay by observing how strenuously Kant denies that neighbor-love can be regarded as a matter of holding or expressing feelings of love, compassion, or sympathy. But now I must add that in several almost casual remarks in *The Doctrine of Virtue,* Kant suggests that it is precisely our duty to try to develop these feelings. One of these remarks occurs in the context of a denial that benevolence be thought of as involving emotional love of neighbor. Contrasting this with active efforts at neighbor-love, Kant observes:

> *Beneficence* is a duty. If someone practices it often and succeeds in realizing his beneficent intention, he eventually comes to love the person he has helped. Hence the saying: "you ought to *love* your neighbor as yourself," does not mean that you ought

> immediately (first) to love him and (afterwards) through the means of this love do good to him. It means, rather, *do good* to your fellow-man, and your beneficence will produce love of man in you (as an aptitude of the inclination to beneficence in general). (MS 203)

Somewhat later, in the context of a discussion of sympathy, Kant gives further concretion to this suggestion:

> But while it is not in itself a duty to share the suffering (as well the joys) of others, it is a duty to sympathize actively in their fate; and to this end it is therefore an indirect duty to cultivate the compassionate natural (esthetic) feelings in us, and to make use of them as so many means to sympathy based on moral principles and the feeling appropriate to them. It is therefore a duty not to avoid places where the poor who lack the most basic necessities are to be found but rather to seek them out, and not to shun sick-rooms or debtors' prisons and so forth in order to avoid sharing painful feelings one may not be able to resist. For this is still one of the impulses that nature has implanted in us to do what the representation of duty alone would not accomplish. (MS 250f.)

Our appreciation of the depth of these remarks is enhanced when we recall that Kant was no friend of a social order based on great inequalities in income and that he occasionally even condemned economic philanthropy because of the injustices on which it depends.[14] Kant would never have us *use* the suffering of others as a means to our moral elevation, but he would call us to an active involvement in whatever unavoidable suffering our

---

[14] In the context of a discussion of casuistical questions connected with beneficence, Kant remarks: "Having the means to practice such beneficence, as depends on the goods of fortune is, for the most part, a result of certain men being favored through the injustice of the government which introduces an inequality of wealth that makes others need their beneficence. Under such circumstances, does a rich man's help to the needy, on which he so readily prides himself as something meritorious, really deserve to be called beneficence at all?" (MS 248).

neighbors experience, both as a means to our moral growth and as an encouragement to continually active neighbor-love.

We therefore come full circle. The philosopher who had initially banished compassion and feeling from the moral life now appears, against the background of a rigorous understanding of concepts, as a proponent of sacrificial altruism and compassionate involvement with the suffering of others. Does this, then, express some of the very central ideas of the Christian conception of *agape?* Perhaps not; there may be specifically religious dimensions of *agape* and possibly also extreme elements of supererogation missed by Kant. Nevertheless, I believe there is ample reason to reaffirm what I said at the outset: those interested in deepening their understanding of the meaning of Christian neighbor-love will not do badly by starting their reading with Kant. His views may not only chasten the excesses to which *agape* has sometimes been subjected, but may also express some of the rational considerations on which Christians have implicitly drawn in trying to make their ethic a practical guide for conduct in society.

3

The Limits of the Ethical in Kierkegaard's *The Concept of Anxiety* and Kant's *Religion within the Limits of Reason Alone*

An individual who turns (or returns) to Kierkegaard's work after a prolonged and sympathetic immersion in Kant's ethical writings has an astonishing experience. At first he encounters what seems to be an alien landscape where the familiar Kantian methods and values—the insistence on reason, rational "objectivity," human autonomy, and supremacy of the moral law—have all been replaced by their opposites—paradox, radical subjectivity, revelation, and faith. His disorientation is accentuated by the constant polemic he encounters against rationalistic philosophy and "speculation," a polemic directed principally against Hegel and his followers but seemingly relevant as well to Kant. Yet, as this Kantian "newcomer" settles in, as he examines his surroundings more closely, he finds that beneath the overgrowth of new forms of expression many key features of the older Kantian landscape are present. Indeed, as he looks further, he discovers that not only are many of the larger features of the intellectual terrain similar, but smaller objects—very particular ideas, patterns of organization, and even concrete illustrations—are often identical. And he wonders whether he has discovered a new region or merely stumbled upon the old after the passage of time.

Nowhere is this sense of familiarity experienced more acutely than in reading Kierkegaard's *Concept of Anxiety*. One who has worked with the more provocative and difficult passages of Kant's philosophy of religion—especially Book One of the *Religion within the Limits of Reason Alone,* where Kant develops his understanding of the "radical evil" of the human will—finds himself immediately at home within the confines of Kierkegaard's penetrating discussion of hereditary sin. Not only are many of Kant's key themes picked up here, but the organization of the two works betrays surprising similarities. True, whole sections of Kierkegaard's discussion are new and contain levels of psychological analysis lacking in Kant. But then, too, some halting suggestions about faith and grace made by Kant are carried by Kierkegaard to conclusions that would have troubled the earlier philosopher. Yet throughout, one has the sense that *The Concept of Anxiety* converses with the *Religion* and, in some senses, represents a natural development of Kant's philosophy of religion. Read together the two treatises even complement one another, with Kant's careful conceptual analysis illuminating Kierkegaard's major unexplained presuppositions and Kierkegaard's psychological insight adding a new dimension to Kant's more formal analysis.

There are several broad parallels between the thinking of Kant and Kierkegaard that should be mentioned at the outset. Some of these parallels have recently been emphasized by commentators who have challenged the conventional wisdom that places Kierkegaard in sharp discontinuity with the previous rationalist tradition.[1] These parallels not only provide further evidence for Kierkegaard's possible indebtedness to Kant, but also furnish much of the intellectual background for both Kant's and Kierkegaard's discussions of sin.

[1] See, for example, Robert L. Perkins, "For Sanity's Sake: Kant, Kierkegaard, and Father Abraham," in *Kierkegaard's* Fear and Trembling: *Critical Appraisals,* ed. Robert L. Perkins (Tuscaloosa: University of Alabama Press, 1981) 43–61; also Jerry H. Gill, "Kantianism," in *Kierkegaard and Great Traditions,* eds. Niels Thulstrup and Marie M. Thulstrup (Copenhagen: C.A. Reitzels Boghandel, 1981) 223–29.

Foremost among the features common to Kant and to Kierkegaard is an anthropology that emphasizes human freedom. Both men view the human being as a limited, finite creature pulled by natural "inclinations" and desires, but, nevertheless, possessing the capacity for controlling impulse and desires through a process of free choice. This capacity fundamentally distinguishes man from other creatures; it confers a special "dignity" on *all* who are human (whatever their stations in life) [2]; and it renders each individual's existence a perpetual striving toward a higher destiny.

For both of these thinkers, the reality of freedom and the importance of this higher destiny are known through an "imperative" laid upon each person's existence. For Kant, this imperative takes the form of reason's moral law (the categorical imperative), while, for Kierkegaard, it is a broader requirement of moral and spiritual development and individuation. Kant and Kierkegaard's understanding of the specific norms bearing upon human life, as well as their respective moral and religious ideals are not identical. Indeed, it has been argued that they are fundamentally different—that Kant's a prioristic, formalistic moral law differs from Kierkegaard's understanding of the normative pattern for the religious individual.[3] I am not convinced that this is true. Individual self-development plays a far more important role in Kant's thinking than is commonly recognized (as evidenced, for example, by his stress in *The Metaphysics of Morals* and the *Lectures on Ethics* on the priority of duties to oneself) (MS 214–42; LE 116ff.). There are also points in Kierkegaard's writings where he displays an appreciation very similar to Kant's of the basic minimal rules of social interaction (which Kierkegaard terms

---

[2] Compare Kant's affirmation of universal human dignity in GR 53f. [435f.]) with Kierkegaard's remarks in EO2 181, 280.

[3] This is the contention of George J. Stack in his *Kierkegaard's Existential Ethics* (Tuscaloosa: University of Alabama Press, 1977) 169f.

the "negative" duties and Kant the "perfect" duties).[4] Important as this matter is, however, in both men's treatment of the problem of sin, the precise content of their respective ideals is less important than their view of the ideal as a demanding challenge before which every individual is likely to fall short.

In addition to agreeing about this sense of requirement, both Kant and Kierkegaard believe that responsibility extends to the exercise of one's will, not to the actual consequences generated by one's choice. Both also seem to agree, however, that in willing, one's responsibility encompasses the *likely* consequences of one's acts. Neither thinker, in other words, appears to advocate a position of benevolent dispositions combined with an attitude of "consequences be damned." In fact, it is the rigor of this requirement imposed upon the will, a responsibility for attending to every *remotely foreseeable* consequence of one's choice, that partly shapes each thinker's perception of the stringency of the moral demand.[5] Nevertheless, in the last analysis it is for the informed exercise of one's will alone that one is morally called to account. This accountability means that for both Kant and Kierkegaard, the forum of moral judgment is essentially internal. Both view moral judgment as first of all an act engaged in by the self against the self. No outer human court is more capable or more demanding than the forum of conscience, and because of the nature of moral judgment, only God (who sees the heart) shares with the self the role of one's accuser and judge.

Mention of God as surveyor of the inner moral life points to a final broad parallel between the thinking of Kant and Kierkegaard: their shared belief that morality must find its

---

[4] Compare, for example, Kant's distinction between "perfect" and "imperfect" duties in *The Doctrine of Virtue* (MS 65, 194) and between "obligatory" and "meritorious" duties in GR 47f. [429f.]) with Kierkegaard's classification of the duties in EO2 267f.

[5] A view of this sort appears to be given especially sharp expression by Kierkegaard in "Guilty?"/"Not Guilty?" (SLW 381).

grounding in religious faith. Those whose familiarity with Kant is limited to *The Foundations of the Metaphysics of Morals,* with its insistence on human autonomy in ethics and its sharp criticism of religious "heteronomy," will find this emphasis on faith markedly un-Kantian. But it is the point of Kant's subsequent writings on the philosophy of religion that religious belief is needed to provide rational coherence for the moral exercise of the will. Kant goes so far as to declare that without faith in a moral God, the moral law itself would become vain and imaginary (KPV 118 [114]; See also KU 120f. [452]). Kant's detailed argumentation here is sometimes called the moral "proof" of God's existence, but it is less a proof than a rigorous exploration of the logical preconditions of rational moral commitment.[6] For Kant, religious faith derives not from idle speculative interests but from an urgent choice made in the sphere of moral life. In this respect for Kant—no less than for Kierkegaard—faith involves an act of will and a leap. It follows from commitment to the moral life. And it is a choice made in defiance of what worldly "wisdom" and experience tell us—for in the world, moral order and moral commitment find little support. It is also a choice always marked by uncertainty. Kierkegaard's existence lived "out on 70,000 fathoms of water" (CUP1 204) has a counterpart in Kant's "wavering" rational faith. For both thinkers, therefore, faith stands in a different epistemological category than knowledge. It always lacks certainty, but far from being a reason for regret, faith's objective uncertainty for both thinkers is a cause for celebration. As a result of its uncertainty, faith becomes not a lethargic and fearful response to God's overwhelming power or presence, but a free and courageous act.[7]

---

[6] For a fuller discussion of the logic of Kant's argument, see my *Religious Reason,* chapters one through three.

[7] Compare Kant's concluding remarks in KPV 152f. with Kierkegaard's comments in CUP1 182, 381f.

It should be clear, then, that many basic tenets of Kierkegaard's thinking are at least prefigured in Kant's work. Of course, there seem to be important differences as well. For all his interest in religion, for example, Kant would limit religious faith and speculation to a zone sternly ruled by morality. In Kant's view, morality always remains supreme and religion is its servant. It follows from this view that for Kant there are no direct duties to God. Full religious obedience can be accomplished by the performance of one's duties to oneself and to one's neighbor, and a religious belief that in the name of divine service contravened morality would be denounced by Kant as a form of "heteronomy." Furthermore, since Kant believed that all the religious beliefs required for moral commitment are freely furnished by one's own reason, revelation and historical faith are for him "in principle" unnecessary.[8] This means concretely that for Kant, Abraham is not the "father of faith," as he is for Kierkegaard, but is a scandalous fanatic best forgotten. As striking as these contrasts seem to be, they serve in a peculiar way as a fitting introduction to the specific parallels to be discussed here between Kant's *Religion* and Kierkegaard's *Concept of Anxiety*. It is in the *Religion* that these two themes that later are to preoccupy Kierkegaard so much receive explicit attention. It is here that Kant discusses extensively the matter of duties to God and concludes that there are "no special duties" of this sort distinguishable from one's duties to self and to neighbor (Rel 142n.). And it is here that Kant first broaches the matter of Abraham's sacrifice of Isaac, contending that no command in violation of morality may be obeyed (Rel 175).[9] Because of the centrality of these themes in the *Religion*, it is reasonable to suppose that if Kierkegaard is not in agreement with Kant on these points he may be involved in dialogue with him. If

[8] Robert L. Perkins, "For Sanity's Sake: Kant, Kierkegaard, and Father Abraham," 52.

[9] Kant deals similarly with Abraham in CF 115n.

one further bears in mind the possibility that Kierkegaard's *Fear and Trembling*—where both the idea of direct duties to God and Abraham's obedience are defended—may be not so much a discussion of heteronomous religious morality as an anguished treatment of the problem of repentance (FT 108f.), then even the disagreement between Kant and Kierkegaard on these matters may be less dramatic than it first appears. As I shall show, Kant also recognized that repentance is the most fearful challenge to our ordinary notion of justice.

With this background, one might now turn to the specific correspondences between the *Religion* and *The Concept of Anxiety.* In particular, four major points deserve mention. They are: (1) the ideality of ethics and the absoluteness of the moral command; (2) the "inexplicability" of sin despite its relationship to human nature; (3) Adam as Everyman through whose sin all humans become corrupt; and (4) the problematic nature of repentance and the necessary role of faith and grace in the economy of redemption.

The Ideality of Ethics

In the introduction to *The Concept of Anxiety,* Kierkegaard seeks to explain why he will devote so much attention to the notion of hereditary sin. He does so by maintaining that with the appearance of sin, "all is lost for ethics" (CA 19). Because the idea of hereditary sin denies the claim that virtue can be realized, it expresses in *dogmatic* form what ethics knows the moment that sin comes forth: that sin and ethics are radically incompatible and that ethics must be "shipwrecked" on sin's appearance.

Behind this strong statement of ethics' incompatibility with sin is a further idea: that ethics is an "ideal" science. "Ethics," says Kierkegaard, points to ideality as a task and assumes that every man possesses the requisite conditions" (CA 16). For this reason, ethics will have nothing to do with the idea that an individual cannot meet its demands or that he can only partly do so. Ethics is, and should be, uncompromising in what it requires:

> The more ideal ethics is, the better. It must not permit itself to be distracted by the babble that it is useless to require the impossible. For even to listen to such talk is unethical and is something for which ethics has neither *time* nor *opportunity.* Ethics will have nothing to do with bargaining. (CA 17)

It is this understanding of ethics' ideality, then, that underlies Kierkegaard's view of sin. Without this ideality, sin would not be the fundamental challenge to ethics that it is, nor would it need to be transcended in religion. But why does Kierkegaard insist that ethics—as the normative "science" bearing on human life—has this ideal character? At least on the surface, this view appears unreasonable. After all, no one is entirely good. We all occasionally fail to perform the moral duties we know are incumbent upon us, but we do not from that conclude that we are "unethical." Nor do we seem to believe that moral judgment requires unerring excellence or that it casts a wholly negative light on partial moral efforts. Yet this is just what Kierkegaard seems to be saying. In a passage of the *Postscript,* he forcefully expresses this view:

> A person can *be both* good and evil, just as it is quite simply said that a human being has a disposition to both good and evil, but one cannot *simultaneously become* good and evil. Esthetically, the poet has been required not to depict these abstract models of virtue or diabolical character but to do as Goethe does, whose characters are both good and evil. And why is this a legitimate requirement? Because we want the poet to depict human beings as they *are,* and every human being *is* both good and evil, and because the poet's medium is the medium of imagination, is being but not becoming, at most it is a very foreshortened perspective. But take the individual out of this medium of imagination, out of this being, and place him in existence—then ethics immediately

> confronts him with its requirement, whether he now deigns to become, and then he becomes—either good or evil.... This *summa summarum*, that all human beings *are* both good and evil, is of no concern at all to ethics, which does not have the medium of *being*, but of *becoming*.... (CUP1 420f.)

Kierkegaard's point here is poetically expressed. His distinction between *being* and *becoming* suggests a more basic distinction between what the lived reality of a person's life has been and what that person *wills* to make himself in the present moment of moral decision. But beyond this, Kierkegaard is not clear. Why does a single moment of willing determine the moral value of a whole course of life? And why is that willing either good or bad? Can there not be an "intermediate" kind of willing, a qualified willing that is less than perfect but that is not immoral?

Kierkegaard does not answer these questions. But Kant discusses them in the *Religion*. Here, too, as a preamble to a discussion of sin, one finds an insistence on the ideality of ethics. But Kant's treatment of this issue—especially when connected with some of his earlier discussion in the *Foundations of the Metaphysics of Morals* and the *Critique of Practical Reason*—helps clarify what Kierkegaard only presumes.

Kant begins his remarks with the question of whether by nature man is good or evil. Very much like Kierkegaard, he considers at the outset the possibility that such absolute evaluations may be inappropriate:

> The conflict between the two hypotheses ... is based on a disjunctive proposition: *Man* is (by nature) *either morally good* or *morally evil*. It might occur to anyone, however, to ask whether this disjunction is valid, and whether some might not assert that man is by nature neither of the two, others, that man is at once both, in some respects good, in other respects evil. Experience actually seems to substantiate the middle ground between the two extremes.

But almost as soon as he raises this possibility, Kant dismisses it:

> It is however, of great consequence to ethics in general to avoid admitting, so long as it is possible, of anything morally intermediate, whether in actions *(adiophora)* [sic] or in human characters; for with such ambiguity all maxims are in danger of forfeiting their precision and stability. (Rel 17f.)

Here, then, is the very ideality that Kierkegaard assumes. But why does Kant hold what he himself concedes is a "rigoristic" view (Rel 17f.). Why not a less precise and more ambiguous "intermediate" position? The roots of the answer to this lie in Kant's basic ethical theory. The morally good person, for Kant, is one whose will is determined not by self love (Kant's very general term for private or particular interests of any sort), but by duty. Duty, in turn, is the dictate of the categorical imperative, that formal principle requiring all maxims (all proposed general rules of conduct; all policies one has set for oneself) to be capable of being willed at the same time as universal law.[10] The morally good person, therefore, is one who chooses to establish this imperative as the supreme guide to his or her conduct. Of course, one need not do this. One can will *never* to obey the categorical imperative (something Kant views as humanly impossible—Rel 30). Or one can will *sometimes*—even in most cases—to obey this imperative, but on occasion to retain the option of not obeying.

This last position would seem to be an intermediate one between good and evil. However, as Kant points out in the second *Critique* and the *Religion,* it really is not intermediate at all. For, a rational person who wills *on occasion* not to obey the moral law must have some higher general principle that determines what he shall do in each instance of choice. But what could this principle

[10] This imperative is usually taken to imply the requirement that maxims be universalizable. For a different interpretation, one stressing the universal *acceptability* of maxims, see my *Religious Reason,* chapter one.

be but the principle of "self love" or private interest? The maxim of the individual who wills occasional obedience to the moral law is one to the effect that he will obey this law *unless* it proves too onerous for him to do so. Since his final referent is to what is convenient to him, this individual allows self-interest to become the arbiter of what he shall do, and he stands with those who make self-interest the first consideration of their will. There is, therefore, no intermediate position between unqualified obedience to the moral law and disobedience. In the clear-sightedness of honest moral analysis, one either is moral or is not. In characteristically difficult terms, Kant spells this out in his discussion in the *Religion*:

> [T]he answer to the question at issue [whether man is by nature good or evil] rests upon the observation, of great importance to morality, that freedom of the will is of a wholly unique nature in that an incentive can determine the will to an action *only so far as the individual has incorporated it into his maxim* (has made it the general rule in accordance with which he will conduct himself); only thus can an incentive, whatever it may be, co-exist with the absolute spontaneity of the will (i.e. freedom). But the moral law, in the judgment of reason, is in itself an incentive, and whoever makes it his maxim is *morally* good. If, now this law does not determine a person's will in the case of an action which has reference to the law, an incentive contrary to it must influence his choice; and since, by hypothesis, this can only happen when a man adopts this incentive (and thereby the deviation from the moral law) into his maxim (in which case he is an evil man) it follows that his disposition in respect to the moral law is never indifferent, never neither good nor evil.
>
> Neither can a man be morally good in some ways and at the same time morally evil in others. His being good in one way means that he has incorporated the moral law into his maxim; were he, therefore, at the same time evil in another way, while his maxim would be universal as based on the moral law of obedience

> to duty, which is essentially single and universal, it would at the same time be only particular; but this is a contradiction. (Rel 19f.)

The insistence on ethics' ideality is not, therefore, just a matter of private moral preferences, nor is it a matter of poetic exaggeration introduced by Kierkegaard to shape his case against ethics. Kant's more formal analysis helps sustain what both he and Kierkegaard assume. (This analysis, incidentally, seems to me not to be a philosopher's abstraction. It has powerful human significance. In the wake of the wars and mass persecutions of this century, we have perhaps learned that those who make only qualified moral commitments ["I will be moral unless ..."] often behave no differently in the crisis of choice than those who are openly wicked. As Kant saw, a contingent commitment to moral duty can often be the same as no commitment at all.)

The "Inexplicability" of Sin Despite Its Relationship to Human Nature

This understanding of the ideality of ethics furnishes one part of the structure of thought that leads both Kant and Kierkegaard back to an appreciation of the religious idea of original sin. Another major component in that structure, however, is their view of human nature. For both Kant and Kierkegaard, man's nature makes sin possible. At the same time, it in no way requires or "causes" man to sin. In fact, both thinkers absolutely repudiate any effort to provide a causal explanation of sin, and both strongly affirm sin's origination in an inherently "inexplicable" act of human freedom.

I have already indicated that Kant and Kierkegaard share an anthropological perspective that sees man as a composite of finitude and transcendence (freedom). For Kierkegaard, sin has its beginning when the self, rendered "dizzy" by the possibility of freedom, finds itself "laying hold of finitude to support itself" (CA 61). While Kant's discussion of sin lacks the psychological dimensions of Kierkegaard's, his moral analysis reveals a broadly similar dynamic. Man is a creature subject to two major "predispositions," Kant tells us. One is the moral predisposition that, if no other incentive worked in opposition to it, would direct the will. But man also has a "natural" predisposition arising from his sensuous nature, the drives and desires he possesses as a finite living being. Were he to subordinate these sensuous impulses to the moral law, he would be virtuous. But in the instant of choice, man somehow inverts the order of priorities. He reaches for the objectives of finitude, and the result is moral evil or sin.

Why does the individual characteristically choose in this fashion? Kant takes pains throughout his discussion in the *Religion* to point out that man's sensuous nature is not to blame. In agreement with the classical Christian tradition—and with Kierkegaard (CA 27, 79; EO2 50)—he maintains that the problem lies not in man's sensuous nature but in his will. To think otherwise would deny man both freedom and culpability (Rel 19f.). Furthermore, Kant repeatedly insists that man's sensuous desires also form the material with which reason's formal moral imperative must work. The virtuous human being is not one *without* desires—for that would be no human being at all—but is rather the individual who subjects whatever desires he has to the constant regulation of the moral law. Man's sensuous nature, therefore, though it provides an occasion for both good and evil, is not the source of evil, in Kant's view (Rel 27, 30, 51). Kant would surely agree with Kierkegaard that moral evil stems not from one part of the synthesis that is man, but from "a misrelation in a relation of a synthesis that relates itself to itself" (SUD 15).

Because he refused to view moral struggle simply as a war between reason and desire, Kant dearly recognized the *rational* possibility of the choice for immorality. In the *Critique of Practical Reason* (115f., 120f.) and again in the *Religion* (50), Kant repeatedly rejects the Greek view that wrongdoing is folly—the overcoming of reason by desire. A human being's reason can be exercised coherently either in a moral or a selfish fashion, Kant believes. No argument to the effect that immoral willing is imply irrational is convincing. Just as natural causes do not require an individual to sin, neither does reason require him to be moral. Kant thus fully perceived and appreciated the depths of human freedom in this regard.

But Kant will not go beyond observing the possibility of sin. He deliberately refuses to give an answer to the question of why an individual should choose the path of immorality. And in this respect, Kierkegaard appears to follow him. Indeed, this refusal to provide an *explanation* of sin is one of the most striking parallels between the *Religion* and the *Concept of Anxiety.* In Kierkegaard's words, sin admits no explanation: "To want to give a logical explanation of the coming of sin into the world," says Kierkegaard, "is a stupidity that only can occur to people who are comically worried about finding an explanation" (CA 50). Sin cannot be explained, he insists because it rests on freedom and like any free act in its very concept it resists antecedent explanation. Sin must thus be thought of as entering the world in a qualitative leap (CA 48, 111), with the "suddenness of the enigmatic" (CA 30).

Kant's agreement here is complete. He resists the effort to explain sin in terms of any "determination" or causation of the will by the natural impulses. Moral evil stems not from these impulses but from a *choice* on our part to allow them sway over the dictates of the moral law. But because this choice rests on freedom, on the reality of our nondetermination by natural causes, it defies any kind of causal explanation and its "ground" or basis must remain forever inscrutable to us. The effort to discover an "explanation" further back beyond the brute fact of a free choice on our part must always re-encounter the mystery of freedom.

> That the ultimate subjective ground of the adoption of moral maxims is inscrutable to us is indeed already evident from this,

> that since this adoption is free, its ground (why, for example, I have chosen an evil and not a good maxim) must not be sought in any natural impulses, but always again in a maxim. Now since this maxim also must have its ground, and since apart from maxims no *determining ground* of free choice can or ought to be adduced, we are referred back endlessly in a series of subjective determining grounds without ever being able to reach the ultimate ground. (Rel 17n.)

Adam as Everyman through Whose Sin We Become Corrupt

Our inability to explain sin means, of course, that wrongdoing is not "necessary." For Kant and for Kierkegaard, sin is sin: imputable wrongdoing and wrongwilling—evil that can and should be avoided. Man's total situation (his natural desires, his finitude, and the ought that bears down on his life) may explain why sin is possible, but it never requires sin. Each human being is constitutionally able to live his life in a proper way without sin. But no one does. For both Kant and Kierkegaard, the misuse of freedom is a universal fact. Not only is every individual guilty of sin—and the most guilty are those who deny this fact—but this sin is profound, enduring, and seemingly inescapable. As has just been seen, the fact of sin cannot be explained. But sin's omnipresence, its persistence, and man's bondage to it are staples of the Christian understanding of biblical faith. In another remarkably similar move at this point in their discussions, both Kant and Kierkegaard turn for deeper insight to the Genesis account of Adam's fall.

The discussions of Adam in the *Religion* and in *The Concept of Anxiety* both begin with the assertion that man's fall cannot be viewed as a singular historical occurrence that subsequently infects all men or causally brings about future sin. Kierkegaard is emphatic on this point. To explain hereditary sin through every man's relationship to Adam and not through the individual's "primitive" relationship to sin, he says, is to place Adam fantastically outside history and outside the human race (CA 26–28). Hereditary sin would then be something present, but Adam would be "the only one in whom it was not found." This interpretation, Kierkegaard maintains, confuses everything. If Adam's sin has any meaning at all, if all humans somehow participate in that sin, a different explanation must be sought. It is to be found, Kierkegaard argues, in the fact that Adam's sin, the first sin, is "something different than *a* sin, (i.e., a sin like many others)" and "something different from *one* sin (i.e. no. 1 in relation to no. 2)" (CA 30). No, it is the first sin because it "constitutes the nature of the quality: the first sin is the sin" (CA 30). In this sense, Adam's sin is not directly causative but is rather prototypical and representative: Adam is at once "himself and the race," Kierkegaard continues, and "that which explains Adam also explains the race and vice versa." Developing this point, Kierkegaard suggests that in Adam's experience we see truths that are daily reaffirmed in each of our lives. These include the truth that sin presupposes itself, that nothing "causes" sin but sin, and that sin enters the world suddenly by a leap (CA 32). Similarly confirmed is the claim that through Adam's *first* sin, sin entered the world. "Precisely in the same way," says Kierkegaard, "it is true of every man's first sin that through it sin comes into the world" (CA 31).

With respect to Adam, Kant almost merely assumes what Kierkegaard feels compelled to argue for at length. Since Kant's basic endeavor in the *Religion* is to demonstrate the rational, moral significance of key tenets of biblical faith, there is never a question for him of interpreting original sin literally as a corruption handed down from our common ancestor. He quickly brushes aside this idea as "inept" (Rel 35) and then turns to a view of Adam as prototypical human being and to the Genesis story as a "representation" of fundamental moral truths. Of Adam he says, *"Mutato nomine de ta fabula narratur"* ("change but the name, of you the tale is told") (Rel 37), a remark reminiscent of Kierkegaard's *unum noris omnes* ("if you know one, you know all") (CA 79).

The Genesis story of Adam's transgression conveys for Kant enduring moral truth. In the narrative of the Fall, for example, is conveyed the understanding that in the search for the origin of evil actions, "every such action must be regarded as though the individual had fallen into it directly from a state of innocence" (Rel 36). This view is necessary because "whatever natural causes may be influencing him, ... his action is yet free and determined by none of these causes; hence it can and must always be judged as an *original* use of his will" (Rel 36). This understanding, he adds, agrees very well with the Genesis account, for there evil does not start from a propensity but as a free transgression of the divine command. Kant does not use the words, but his point is very similar to Kierkegaard's: sin presupposes itself.

Kierkegaard's individualized interpretation of Adam's "first" sin as the cause of sin—his claim that sin comes into the world through *every man's* first sin—also has powerful resonance in Kant's discussion. Like Kierkegaard, Kant individualizes the historical element in accounts of Adam's fall and contends that its significance is to be found in the fact that every one of our own transgressions has a quality of pointing backward to a previous time in our lives and a previous transgression. As we trace this sequence backward, we find sin commencing in the very earliest exercise of our will (Rel 38).

This "individualized" interpretation of the "first" sin is important, not just because it seems to be shared in part by Kant and Kierkegaard, but because it may also help to explain a far deeper correspondence between the two men: their conviction of the truth of the Christian understanding of man's total and seemingly inescapable bondage to sin. Neither thinker, of course, will accept a purely historical explanation of this understanding.

For Kant, in particular, it is preposterous to predicate sin of the species in such a way that it becomes part of a human inherited constitution, for that would be contrary to the very idea of freedom. Sin would no longer be sin. Yet Kant does develop an understanding of the "first" sin such that it becomes for each person a cause of virtually inescapable bondage to sin and of total, limitless guilt. Although there is no direct parallel to Kant's analysis in *The Concept of Anxiety,* Kierkegaard does share Kant's conclusions here, and this may be one of those points where Kierkegaard stands in unacknowledged debt to his predecessor, who has managed to refurbish a difficult dogmatic conception.

Two major considerations underlie Kant's discussion of the virtual inescapability of sin. One has already been touched on: the ideality of the moral command, the requirement that to be morally good an individual must will unwavering obedience to the dictates of duty. The second consideration is that no individual is in a position to certify (to himself or to others) that he is morally good. This kind of positive self-evaluation would require the individual to affirm that he has resolved *always* to obey the moral law. Such resolve must be genuine; not a passing whim, but a genuine inner conviction buttressed by a consistent pattern of moral willing.

But can any individual so positively evaluate his will when he subjects it to an honest and searching scrutiny? Kant thinks not, and he identifies three deeply interrelated reasons why this is so. First, there is the problem that even the seemingly firmest present resolve is not yet equivalent to a righteous course of life (Rel 60). Knowing what he does about himself and about human beings generally, the well-intentioned individual must naturally be reluctant to base a globally positive self-estimate on a single moment of willing, when what is required for virtue is a *life* of such willing. A second problem exacerbates the first: the notorious changeableness of the human will. In view of the will's fickleness, what gives an individual the confidence to believe that his present resolve will hold? Kant maintains that while it may seem morbid to dwell on this inconstancy, it is the honest and morally responsible thing to do. Better to approach one's self-justification with "fear and trembling," he contends, than to hold, as we are all prone to do, an exceedingly good opinion of oneself (Rel 62).

Finally, there is the weightiest problem of all: the enormous and seemingly insuperable burden of one's *past* defections from the moral law—even the burden of that single "first" sin in the remote beginnings of the use of one's will (Rel 65). As we survey our past, Kant suggests, each of us can identify moments when we knowingly and freely subverted the priority of duty or where we are at least uncertain as to which motive took precedence in our choice. Past moments of this sort, however, cast a dark shadow on our present and our future. In part, this is the problem of making amends for these previous wrongs—the problem of repentance. But there is also the problem of what these acts say about the very character of our will. If these acts had not been free, if they had been conditioned by forces beyond our control, or if they had been shaped by ignorance, we could presumably put them behind us and embark upon our present life of free, moral choice. But if we freely and knowingly made vicious choices in the past, what conceivably can prevent us from doing so again? Precisely because they are free, moral choices thus have a "timeless" quality. One act has enduring significance as a measure of how we use our will, and one past act remains a persistent, recurrent possibility for the future.

All these considerations lead Kant to conclude that every honest person must properly regard himself as morally deficient. Indeed, scrupulous inward self-examination that will not content itself with mere outward show or partial success, and the stern ideality of the law that recognizes no intermediate between good and vicious willing—these combine to point to the conclusion that we are all possessed of a radically evil will. Moreover, since this evil traces back to the first or original exercise of our wills, it is properly thought of as a form of original sin. Kant expresses this conclusion in a passage where he tries to connect his analysis to the insights of biblical and Christian faith:

> This debt which is original, or prior to all the good a man may do—this, and no more, is what we referred to in Book One as the *radical* evil in man—this debt can never be discharged by another person, so far as we can judge according to the justice of our human reason.... Now this moral evil (transgression of the moral law, called SIN when the law is regarded *as a divine command*) brings with it endless violations of the law and so *infinite* guilt. The extent of this guilt is due not so much to the *infinitude* of the Supreme Lawgiver whose authority is thereby violated ... as to the fact that this moral evil lies in the *disposition* and the maxims in general, in *universal basic principles* rather than in particular transgressions. (The case is different before a human court of justice, for such a court attends merely to single offenses and therefore to the deed itself and what is relative thereto, and not to the general disposition.) It would seem to follow, then, that because of this infinite guilt all mankind must look forward to *endless punishment* and exclusion from the kingdom of God. (Rel 66)

It needs hardly be said how thoroughly Kierkegaard agrees with Kant at this point. Indeed, a passage in *The Concept of Anxiety* expresses these ideas in terms so similar to Kant's that they suggest direct borrowing.

> Whoever learns to know his guilt only from the finite is lost in the finite, and finitely the question of whether a man is guilty cannot be determined except in an external, juridical and most imperfect sense. Whoever learns to know his guilt only by analogy to judgments of the police court and supreme court never really understands that he is guilty, for if a man is guilty, he is infinitely guilty. (CA 161)

The Problematic Nature of Repentance and the Role of Faith and Grace in the Economy of Redemption

The course of Kant's analysis takes him to the familiar impasse identified by Christian theology: the seeming embeddedness of sin in human nature. Quite unexpectedly for an Enlightenment thinker, Kant has shown the enormous difficulty of becoming a virtuous person. A rationalist, he has shown the limits and difficulties of rationalism. But Kant does not stop there; he seeks to point a way out of the deep trough of iniquity he has discovered. That path lies through repentance, a topic to which Kant gives considerable attention in the *Religion.* Kant's position here is that it is possible for humans to exercise their will to repent and by so doing to put sin behind them as they embark on a new, upright moral life. In this respect, his position would seem to differ from Kierkegaard's. While Kierkegaard's understanding of repentance is not always clear, it seems to be his principal view that repentance does not alleviate but only sharpens the problem of sin. Repentance, he says repeatedly, is at once the "highest ethical expression" but also "the deepest ethical contradiction" (CA 117; FT 108). This characterization seems to be true, on the one hand, because repentance is a distraction from moral conduct. On the other hand, repentance, even if ethically allowable, is impotent before sin: "repentance cannot annul sin; it can only sorrow over it" (CA 115; FT 109). For these and other reasons, full redemption depends upon a direct relationship with God and is the product of faith and grace.

Before one concludes that there is complete disagreement between these two thinkers on this point, one must look more closely at Kant's discussion of repentance. In doing so, I think one can see that because of the great honesty and penetration of his analysis, Kant partly subverts his own point. Rather than demonstrating that repentance is a remedy for sin, Kant shows almost the reverse: that repentance is powerless before sin and antagonistic to some key requirements of ethics. In this respect, a careful reading of the discussion must have been enormously instructive to a thinker like Kierkegaard, who doubts man's ability to effect his own redemption through moral striving. Kierkegaard's repeated allusions to repentance as the "shipwreck" of ethics, in other words, find solid, if unintended, support within Kant's careful discussion in the *Religion.*

The problem that forms Kant's point of departure for his treatment of repentance is that past sins cannot morally be escaped. Each person knows, says Kant, that whatever he may have done in the way of adopting a good disposition, *"he nevertheless started from evil* and this debt he can by no possibility wipe out"(Rel 66). Several different problems follow from this one: First, a person "cannot regard the fact that he incurs no new debts subsequent to his change of heart as equivalent to having discharged his old ones." These prior deeds remain, in Kant's view, an offense that somehow must be offset. Second, no one is able through future conduct to produce a surplus to repay these debts "for it is always his duty to do all the good that lies in his power" (Rel 66).

How, then, can a person make amends? Part of the answer, says Kant, lies in viewing the suffering created in and through genuine repentance as a suitable punishment for these previous wrongs. But this view produces a new and still more serious problem. In the change of heart effected by repentance, a person's former self—the person who performed these wicked deeds—has vanished and has been replaced by a "new person"—one who wills only upright conduct. How, in justice, can punishment be inflicted on the new self who in reality is morally "another person"?

Kant's answer is as complex as the problem. Since the infliction of punishment can be conceived as taking place *"neither before nor after* the change of heart, and is yet necessary," he says, "we must think of it as carried out *during* the change of heart itself and adapted thereto" (Rel 67). Just recompense thus becomes for the repentant the profound sorrow over previous sin experienced during repentance and a humble willingness at that moment to accept all future suffering in life, which, though properly not due his changed person, he nevertheless takes upon himself.[11]

Despite its penetration, Kant's solution to the problem of sin—his view of the adequacy of repentance—is not entirely persuasive. One serious difficulty is how one person's suffering can be taken in repayment of another's moral debt. Even if one concedes that a new moral self can somehow repay the previous self's wrong (an issue tangled in notions of retributive punishment), can suffering justly be inflicted on an "innocent" person? If a person has really repented, it may be that he can view his remorse *during* this change as justified punishment for his other, immoral self (who, for a confused instant, he continues to be). But is it not wrong that he should have to suffer subsequent penalties, as Kant claims? Such suffering inflicted on a good person violates one's sense of justice, and the very idea may cause us to resist the full dynamic of repentance that Kant believes must take place. We are thus led to a kind of wrathful anger at repentance and a hardening in sin of the sort that Kierkegaard explores (CA 116). More serious, however, is the problem of whether *any* self-established and self-accepted punishment can be regarded as adequate recompense for the moral wrong a person has done. Kant has said, after all, that against the ideality of the

---

[11] In the passages that follow these remarks, Kant ingeniously interprets Christ's vicarious atonement as a representation of the essentially vicarious nature of all penitent suffering (Rel 68f.). This interpretation suggests a further parallel between Kant and Kierkegaard: a shared conviction that the truly Christian life is one characterized by the willing acceptance of suffering.

ethical demand, everyone's guilt is infinite. But if this is so, can anything short of total self-annihilation (moral and spiritual death) make repayment for sin? Dare a person regard the remorse experienced during repentance and any subsequent misfortune as sufficient compensation for his wrong—and if he does so has he not precisely missed the full seriousness of his prior conduct? Paradoxically, the most repentant individual must know how inadequate his repentance is to the demand, while only a shallow conscience can regard repentance as a solution to the problem.

Kant's entire discussion, then, seems to point beyond itself. As a means of moral self-justification, repentance is filled with problems. At the same time, however, it readily gives rise to certain religious ideas that supplement its deficiencies. For example, if the self cannot esteem any suffering it undergoes as adequate repayment for its wrongs, it can hope that this suffering may be regarded as sufficient in the eyes of a more knowledgeable, more objective judge. God plays this role for the religious believer. Similarly, if the self cannot certify its will to be reformed, it can hope that God, who sees the heart and who knows both past and future, will find this changed willing as the beginning of a new personhood. The self may even hope that its resolve will somehow be assisted by God in the future. Throughout the *Religion* Kant actually explores ideas of this sort. For example, to understand how a change of heart can come about, he says, one may require a notion of divine grace (Rel 43, 60, 82). In some passages, he even perceives a role for active, divine assistance in the process whereby sin is canceled and a new upright will replaces the old (Rel 159, 179). These passages, however, alternate with others (Rel 19, 55, 134, 179) that point to the moral difficulties of these ideas—for how, Kant asks, can we be judged morally worthy for what we have not done? Throughout, Kant also insists on the abiding priority of our always first reorienting our will toward the moral law.

As other commentators have noted, in the pages of the *Religion* one sees a great philosopher torn between his own best insights.[12] On one side stands an ethical theory whose premise is man's ability and obligation to perform the moral law. On the other side stands an equally powerful series of ideas spawned by that same ethical theory, including an understanding of the depths of human freedom such that even immoral willing is a rational possibility, a sense of the law's stringent demand before which all stand unjustified, and a corresponding understanding of man's virtual bondage to sin. If the *Religion* repeatedly vacillates between an Enlightenment stress on duty and a more orthodox intimation of the need for grace and faith as a way of completing the moral enterprise, it is because Kant was himself divided on these matters. On the one hand, he was the quintessential Enlightenment philosopher. On the other hand, he was a thinker whose very intellectual power shattered some of the confidences upon which Enlightenment thought was built.

With Kierkegaard one enters another world. His intellectual context, his biography, and his temperament make him fully willing to develop the logic of faith. The difficulties in ethics that were for Kant a source of embarrassment become, for Kierkegaard, opportunities. Where Kant was perhaps reluctant to develop some of his best insights, Kierkegaard is eager to do so, and Kant's rigorous analysis appears to provide all the occasion he needs.

Everything said here suggests that in Kierkegaard, Kant found one of the best readers of his *Religion within the Limits of Reason Alone.* Not only does Kierkegaard in *The Concept of Anxiety* (and elsewhere in his writings) appear to draw upon key insights

---

[12] See, for example, the account of Kant's discussion of grace offered by John R. Silber in his introductory essay, "The Ethical Significance of Kant's Religion" in *Religion within the Limits of Reason Alone,* trans. Theodore M. Green and Hoyt H. Hudson (New York: Harper & Row, 1960) cxxxiIlf. For a more sympathetic view see Allen W. Wood, *Kant's Moral Religion* (Ithaca, New York: Cornell University Press, 1970) 208–248.

presented in the *Religion,* but, in some ways, Kierkegaard manages to carry these insights out to the conclusion that Kant, hampered by his own intellectual baggage, was unable to reach.

Did Kierkegaard, then, actually study the *Religion* and draw upon it in writing *The Concept of Anxiety?* This question has been left for last because presently I am unable to answer it. What evidence exists is conflicting and only circumstantial. On one side of the question is the fact that none of the standard biographies mentions the extent of Kierkegaard's familiarity with Kant, and nowhere in his papers or writings does Kierkegaard himself explicitly mention the *Religion.*[13] Then, too, there is the fact that at least some of the points explored here—for example, the moral difficulties of repentance—are touched on by thinkers such as Johann Fichte and Hegel, by whom Kierkegaard was clearly influenced.[14] This connection may suggest a debt to these two nearer figures rather than to Kant. On the other side of the question, there is the fact that Kierkegaard does, at least once, claim familiarity with "Kantian" ethics (EO2 327), and he does refer explicitly to Kant's idea of "radical evil," the central concept in the section of the *Religion* explored here (JP3—3089). The

---

[13] Kierkegaard does refer to several of Kant's writings. For a listing of these references see JP2 611f.; see also Jerry H. Gill, "Kantianism," 228.

[14] Many of the themes touched upon here resonate within the writings of Fichte and Hegel. For example, Fichte's *Attempt at a Critique of All Revelation,* trans. Garrett Green (Cambridge: Cambridge University Press, 1978) broadly shares Kant's conception of the ideality of morality and is critical of nonmoral or purely "historical" elements in Christian teaching. The section on "Absolute Spirit" in Hegel's *Encyclopedia of Philosophy,* trans. Gustav Emil Mueller (New York: Philosophical Library, 1959) bears some similarities to Kant's discussion of sin and repentance. Hegel's treatment of Adam as a representative type in the *Lectures on the Philosophy of Religion,* trans. E.B. Speirs and J.B. Sanderson (London: Kegan Paul, Trench, Trubner, 1895) especially 1:275–79 and 2:202–204, also displays similarities to the work of Kant and Kierkegaard. But in all of these cases, the resemblances are often faint and the texts are only reminiscent of the more full-bodied Kantian arguments on which they are based.

possibility of a debt to Fichte or Hegel on the points discussed is also somewhat doubtful because both thinkers only mention these issues in passing, and one finds in their work none of the careful analysis that makes Kant's treatment so authoritative.[15]

As is, I must leave this historical and biographical question to others for the time being. What I have tried to do here is at least indicate some major continuities between Kant's *Religion within the Limits of Reason Alone* and Kierkegaard's *Concept of Anxiety.* By pointing these out, I hope to have shed new light on some of the important presuppositions of Kierkegaard's work. I also hope to have indicated a relatively unexplored path—one leading through Kierkegaard—by which Kant's philosophy may have influenced subsequent religious thought.

---

[15] For a full accounting of Kierkegaard's reading in Kant, see *Kierkegaard and Kant, The Hidden Debt.*

4

Kant and Kierkegaard on the Need for a Historical Faith: An Imaginary Dialogue

The day is 1 January 2027. Immanuel Kant and Søren Kierkegaard unexpectedly find themselves sharing a corner in the Delta Medallion Club at Denver Airport. Snow has delayed their flights for several hours.

The two great scholars of philosophy and religion are familiar with one another's writings. Thanks to the work of Gill, Glenn, Mehl, Perkins, Green, and others in the last quarter of the twentieth century, it is now well known that Kierkegaard constructed much of his thinking on the foundation of Kant's philosophy.[1] *Kant's familiarity with Kierkegaard is*

---

[1] Jerry H. Gill, "Kant, Kierkegaard and Religious Knowledge," *Philosophy and Phenomenological Research* 28 (1967–68): 188–204; see also his article on "Kantianism," in *Kierkegaard and Great Traditions*; John D. Glenn, Jr., "Kierkegaard's Ethical Philosophy," *Southwestern Journal of Philosophy* 5 (Spring 1974): 121–128; Peter J. Mehl, "Kierkegaard and the Relativist Challenge to Practical Philosophy," *Journal of Religious Ethics* 14, no. 2 (1987): 247–278; Robert L. Perkins, "For Sanity's Sake: Kant, Kierkegaard, and Father Abraham," in *Kierkegaard's* Fear and Trembling: *Critical Appraisals*, ed. Robert L. Perkins (Tuscaloosa: University of Alabama Press, 1981); Ronald M. Green, *Kierkegaard and Kant: The Hidden Debt* (Albany: State University of New York Press, 1992). See Also R.Z. Friedman, "Kierkegaard: First Existentialist or Last Kantian?" *Religious Studies* 18, no. 2 (June 1982): 159–170; Jeremy D.B. Walker, *To Will One Thing: Reflections on Kierkegaard's 'Purity of Heart'* (Montreal, Canada: McGill-Queen's University Press, 1972); also his *Kierkegaard's Descent into God* (Kingston: McGill-Queen's University Press, 1985); Alastair Hannay, *Kierkegaard* (London: Routledge and Kegan Paul, 1982); William D. Peck "On

more recent—one aspect of the education program he has gone through in the decade since his "reanimation" was accomplished by means of modern genetic science and computer technology.

A waiter takes their order and returns with two glasses of California Chardonnay wine and a bowl of Goldfish crackers.

KANT: I really enjoy these little crackers, don't you?

KIERKEGAARD: One of the best of the modern "inventions"!

Recognizing that they have an opportunity to pursue a topic dear to them, the two thinkers quickly turn to ethics and its relationship to religion.

KIERKEGAARD: Let me say, first of all, how pleased I am to have this chance to express my thanks to you. As you know, during my lifetime I was unable to be as public as I would have liked about my reliance on your work. The Danish attitude toward rationalism in general and to Kantianism in particular was so negative in my day. I never joined this chorus of voices. I took a strong stand of opposition to the Hegelians who made light of your work, and I took pains to insert brief but very positive things about your philosophy and your character in my writings.[2] But I admit I never credited you properly.

KANT: I understand. Frankly. I'm not sure what I would have done in your place. As you know, despite my reputation for moral rigor, I made my own compromises with strict veracity. One of these was my pledge to King Frederick to obey his edict not to

---

Autonomy: The Primacy of the Subject in Kant and Kierkegaard" (PhD thesis, Yale University, 1974); C. Stephen Evans, *Subjectivity and Religious Belief: An Historical, Critical Study* (Washington DC: University Press of America, 1982); Geoffrey Clive, "The Connection between Ethics and Religion in Kant, Kierkegaard and F.H. Bradley" (PhD thesis. Harvard University, 1953).

[2] See, for example, PAP VIII2 A 358 n.d.; VIII2 B 81 n.d. 1847 [JP1-649]: cf. XI A 666. 1849 [JP3-3558]; X1, 666 n.d . 1849; X2 A 501 1850.

publish *Religion within the Limits of Reason Alone.* Without saying so, I interpreted this as a pledge to Frederick just so long as he remained alive. Many have since remarked that I was a bit casuistic in my reasoning.

KIERKEGAARD: Since you raise the subject of the *Religion,* let me say here how much this book influenced me. During my student years it was a ray of light in the darkness. Here was the undisputed moral rationalist, the father of the modern concept of moral autonomy affirming the "radical evil" in human beings and our need for divine grace to achieve moral fulfillment!

KANT: It's interesting that you should say that. I confess I was initially unhappy with my conclusions in the *Religion.* I thought I had said all there is to say about rational religious beliefs when I wrote the *Critique of Practical Reason.* A moral governor of the universe, the possible continuance of our life beyond death to accomplish our perfection in moral virtue—I honestly believed these were the only religious concepts we needed to complete the moral life.

KIERKEGAARD: What was it that changed your mind? You know many have said the *Religion* was nothing more than your effort to pacify the orthodox, including your manservant Lampe.

KANT: That's amusing, and unfair to Lampe. He was no fool and disliked priestcraft as much as I did. In fact, the *Religion* was really a surprising consequence of an idea already present in the second *Critique.* It was an idea I initially resisted because I feared its practical implications. I yielded to it only when I became convinced of its truth and power.[3]

---

[3] In offering this view of the relative novelty of Kant's efforts in *Religion* to explore the extent of our rational freedom from morality, I disagree with Dennis Savage's estimate that "Kant's theory of radical evil in the *Religion* contains nothing basically new as compared with his theory of moral good and evil presented in his [earlier) ethical works." See "Kant's Rejection of Divine Revelation and His Theory of Radical Evil," in *Kant's Philosophy of Religion Reconsidered,* ed. Philip J. Rossi and Michael J. Wreen (Bloomington:

KIERKEGAARD: What is that idea?

KANT: That in relation to morality we are radically free. In the second *Critique* I began to explore a basic problem in rational moral justification. We know we are bound by the moral law whose voice is commanding. We also know we are "creatures of needs" who are compelled to seek the satisfaction of our desires, the sum of which constitutes our happiness (GR 21 [405]; KPV 25). Indeed, one purpose of morality is to create the rule of law that permits all persons the ordered pursuit of their happiness. But the question is, how are we to reason when the dictates of the moral law run counter to our valid rational concern with our well being?

KIERKEGAARD: One would think you would reply that we must obey the dictates of the moral law that by its nature requires the subordination of individual ends to the common good.

KANT: That's certainly right. But the question is, how do we rationally justify the priority of moral reason? We can't appeal to the individual's happiness in this context, since this is just what he or she is being asked to subordinate. Nor can we appeal to the satisfaction that comes with virtue (what I termed "self-contentment" in the second *Critique)*. The people who choose to act virtuously certainly experience this, but this is because they have already chosen to give prime importance to their moral self-estimate. The question is, why should they do this? Finally, we cannot argue that objective self-disregarding reason (what I call "pure practical reason") dictates this priority, because it is just this pure reason whose supreme authority is being questioned. Indeed, this may be the only instance in all of rational justification where the authority of pure reason can be impugned. In other words, we find that in this situation all rational justification runs in circles.

KIERKEGAARD: So you are saying we cannot be compelled by reason to accept the priority of the moral law?

Indiana University Press. 1991) 73.

KANT: Exactly. This doesn't mean we can avoid the command of moral reason. Certainly we can never unequivocally justify anyone's unfettered pursuit of personal happiness. Such a policy is insanity and would soon defeat itself. But the priority of moral reason nevertheless defies strict rational justification. To eliminate this problem, I argued in the second *Critique* that our practical reason leads us to entertain certain religious beliefs. To the extent we believe the world may be ruled by an all-powerful and just moral governor who unerringly rewards our virtue (and punishes our vice), we have a reasoned basis for always giving priority to our morally commanding reason.

KIERKEGAARD: You are not saying that morality requires us to hold these beliefs?

KANT: No, not at all. That would be to find rational necessity where, as I have said, none exists. These beliefs are a way of holding together all the conflicting dictates of our practical reason. If we wish to make our reason harmonious with itself in its theoretical and various practical employments (KU 187), we can act morally and subscribe to the religious beliefs that assist us in doing this. But, as I put it, such a position is a choice, "a voluntary decision of our judgment ... itself not commanded" by reason (KPV 146). Hence, our reason permits us to adopt morality and its attendant religious beliefs, but it also permits us, if we are willing to live with conflict at the core of our reason, to abandon morality and these same beliefs.

KIERKEGAARD: I take it that this is the truth in the traditional Christian teaching that wickedness begins with unbelief.

KANT: Precisely. And it is also the source of the radical freedom I spoke of. This is not just a freedom to do as we wish. We have that freedom in any case if we are prepared to act irrationally. But here we have a freedom to act immorally in a way that cannot be rationally condemned. We have a rational freedom for immoral choice.

KIERKEGAARD: And this is the idea that you resisted until you turned to it in the *Religion?*

KANT: That's right. You might say that in the second *Critique* I devoted myself to developing and defending morality's associated religious beliefs. But I failed to dwell on the implications of the sheer voluntariness of these beliefs and the depth of our rational freedom in this area. In the *Religion* the implications of this voluntariness were moved to the fore.

KIERKEGAARD: Can you briefly spell those implications out?

KANT: They proceed in sequence. First, since we are not required by our reason to give priority to the moral law, there is every reason to believe that on occasion we will fail to do so. Second is the moral requirement that we must guarantee unerring obedience to the moral law if we are to claim any moral worth for ourselves at all. The third implication follows from these two: since we can and will fail to uphold the priority of the moral law, we cannot sustain any claim to moral worth.

KIERKEGAARD: Can you please explain your second point?

KANT: Our need to guarantee unerring obedience to the moral law?

KIERKEGAARD: Yes.

KANT: Very simple. We must guarantee unerring right willing because moral worth is an all or nothing affair. One cannot be just a "little bit" immoral. Even one act of wrongwilling evidences a fundamentally immoral disposition, one that places other considerations before the moral law. This is what I called a corrupt "general maxim" of the will. If I may use a metaphor from an area I have just learned about, a person who occasionally subordinates morality to self-interest is like a computer that occasionally miscalculates sums. Such unreliability makes both the person and the machine worthless. This is why in the *Religion* I termed this ever-present tendency to such an inversion of motives "the radical evil" of the human will, an evil that lies at the very root of our morally legislative disposition and that corrupts us fundamentally. In religious terms if we think of our duties as commanded by a morally legislating God, this evil is sin.

KIERKEGAARD: But you are not saying that we must sin?

KANT: No. Absolutely not. This is a matter of free choice, a choice that is even less required by reason than the choice of morality. But it is a rationally possible choice. And what is possible may become actual. This possibility is what weakens our confidence in any judgment of our moral worth.

KIERKEGAARD: Does the problem stem from the fact that we possess many personal desires?

KANT: Yes and no. We are creatures of needs, pulled by inclinations that occasionally war with our moral reason. Without such inclinations, there would be no motive to disobey morality. We would then have a holy will like God. But these inclinations do not themselves explain our wrongdoing. We always know ourselves to be free to resist them and they can sometimes even lead us to virtue (Rel 51–52). Nor does the problem arise just because we face stress, hardship, or want. Inexplicably, we choose to invert the priority due morality in good times as in bad. This is why Scripture presents the fall of man as occurring once, inexplicably and without necessity, but in a way that foreshadows the recurrent fall of all who follow. For if even one of us succumbs to this misuse of freedom, who can confidently assert that he or she will not also do so?

KIERKEGAARD: This, of course, is the philosophical reinterpretation of the doctrine of original sin that you present in the *Religion*.

KANT: Yes. It is the first of several such philosophical reinterpretations of orthodox teachings that I endeavor there. But I want to stress that I'm not looking to the historical fact of Adam's sin. The explanation of sin as an inheritance from our first parents is the most inept one I can imagine (Rel 35). No person can be imputable for the wrongful deeds of another. Adam is each one of us. Experience teaches that at some point we each "fall" freely into the choice of immoral conduct, and even one such fall calls fatally into question the constancy and worth of our moral disposition, convicting us in our own eyes of sin.

KIERKEGAARD: This is an ingenious argument. I've pushed you because I wanted to hear your own synopsis of insights that have had a great impact on my thinking. As you may know, I drew heavily on your arguments to ground my repeated assertion that philosophical ethics leads to its own undoing. For example, in *The Concept of Anxiety* I said: "Ethics points to ideality as a task and assumes that every man possesses the requisite conditions. Thus ethics develops a contradiction, inasmuch as it makes clear both the difficulty and the impossibility" (CA 16–17). Or, again, "An ethics that ignores sin is a completely futile discipline, but if it affirms sin, then it has *eo ipso* exceeded itself" (FT 99). In all this, it was your development of the ideality of ethics, the rigor of the moral demand, and the inevitable but imputable fact of moral failure that informed my thinking.

KANT: Yes, but we must be careful. In reading your work, I feel you wanted to go much farther than I was willing to go.

KIERKEGAARD: Why do you say that?

KANT: Because you use these ideas as a springboard to defend orthodox Christian teaching, especially faith in a historical savior. But just as I vehemently deny that it does us any good to look at the Fall in historical terms, so I deny any ultimate importance to historical events or revelations in the process of our moral redemption. Everything we need for our moral salvation resides within us as a part of our practical reason.

The conversation is *interrupted* as *a waiter asks whether they wish another drink. He tells them the weather has lifted and flights will probably be resuming in the next hour or two.*

KIERKEGAARD: Well, then, it seems our time is short. This difference between us is so important that I would like to focus on it immediately. Let me say that I simply do not understand your position. It seems to me in clear contradiction to almost everything you say.

KANT: What do you mean?

KIERKEGAARD: As I understand your argument in the *Religion*, you frankly acknowledge that we must accept the conclusion that we bear "infinite guilt" for our defection from the moral law (Rel 66)?

KANT: Yes.

KIERKEGAARD: And as a consequence, we merit endless punishment (Rel 66)?

KANT: That's right.

KIERKEGAARD: You further concede that there is a substantial place for divine grace in the process of our moral redemption—that when we reach the depths of our moral self-esteem, we are driven to the possibility of grace as the sole way of escaping moral despair and rationally resuming our moral striving?

KANT: I'm not sure I'm comfortable with the phrase "sole way." The sole way we can regain our lost moral course is to rededicate ourselves to upright willing. We cannot look to anyone else to do our moral work for us. Let me add here that, according to human reason, it does not make sense to believe our moral debt can be discharged by another person, even if that person is declared to be the Son of God. Moral evil is not a transmissible liability that can be made over to another like some commercial debt (Rel 66).

KIERKEGAARD: I understand your insistence on this point, and I might say that I am not a proponent of the scholastic-dogmatic view of atonement. We must ourselves suffer alone. But God's grace can work in us through Christ in other ways besides this kind of substitutionary atonement. The problem is that divine activity seems to have no real place in your scheme. Where does grace fit, if we can achieve moral conversion on our own? I'm sure you've become familiar with modern writers who perceive a deep incoherence in your views at this point. Gordon Michalson, for example, believes that your perspective "wobbles" between two discreet and incompatible traditions of thought: your orthodox past and an Enlightenment perspective.[4] Nicholas Wolterstorff and others maintain that, on the one hand, you recognize the depths of our moral incapacity to achieve enduring rectitude and our need to rely on a power beyond ourselves, but that, on the other hand, you refuse to relinquish a stoic insistence on moral imputability and responsibility.[5] Hence, the contradictions in your arguments and your reluctance to look to saving grace to forgive our sins and redeem our will.

KANT: But I *do* look to saving grace. Michalson and these other modern writers totally miss my point here. I fully admit a role for God's involvement in our moral life through divine grace. First, I appeal to God's timeless intuition to ground our hope that our individual acts of renewed moral willing are in fact part of a course of unvarying rectitude. Second. I look to God for the confidence that our new, upright disposition will remain constant, and I regard this very disposition, which I call our Comforter or Paraclete, as a sign of God's support (Rel 65). Third, though I may not have made this point very clearly, I argue that we may also look to God to accept the penitent suffering we undergo during our moral conversion as adequate to repay the infinite wrong we have done, provided this suffering is pictured as a death endured

---

[4] Gordon E. Michalson, Jr., *Fallen Freedom: Kant on Radical Evil and Moral Regeneration* (Cambridge: Cambridge University Press, 1990) 128, 133.

[5] Nicholas P. Wolterstorff, "Conundrums in Kant's Rational Religion," in *Kant's Philosophy of Religion Reconsidered*, 48–49; John E. Hare, *The Moral Gap: Kantian Ethics, Human Limits, and God's Assistance* (Oxford: Clarendon Press, 1996) 60–62.

once for all by the representative of mankind. This is the proper place for a concept of vicarious suffering, as arising out of our own moral concepts and reflecting the suffering the new, morally reformed person undergoes in leaving behind the old, morally corrupted self (Rel 68).

KIERKEGAARD: Let me say that I find your rationalist interpretation of vicarious suffering very interesting.

KANT: Thank you. I regard it as one of my more penetrating deductions of a concept. But let me make clear how important the reliance on God is even in this rationalized conception. We must hope that our suffering will satisfy our moral debt. We cannot make this judgment ourselves without appearing to escape a merited punishment. In contrast, a moral governor of the universe who truly knows our frailty and our place in the larger moral order can judge us differently than we must judge ourselves (Rel 131n.). Hence, grace (or what classical theologians might call "God's righteousness") permeates my account and is essential to it.

KIERKEGAARD: And yet you still will not relinquish the insistence on individual moral reform and rededication as a first step in this process? Doesn't that return us to the critics who claim you place the emphasis on autonomous human willing rather than God's prior initiative on our behalf?

KANT: Absolutely not. Don't tell me, Kierkegaard, that you, too, fail to see my point here?

KIERKEGAARD: No, in fact, I think I fully understand it. My disagreement with you lies elsewhere and has to do with the role of the historical savior. But I'm trying to become very clear about precisely what you are saying. Let me sharpen the question: How do you reply to the claim that your whole account focusing on human moral willing and rededication (supplemented by grace) seems opposed to a traditional conception of grace, whereby we are first accepted and revived from moral death and only then empowered to accomplish moral rededication?

KANT: I see no conflict between these two accounts. They are one and the same. Whether grace reanimates the will or willing exhibits grace is all the same. In both cases the emphasis must be on *our* willing. Through this lens we mortals perceive divine effects. More than this we cannot know and cannot say. As I observe over and over again in my writings, we have no *knowledge* of noumenal things. We cannot understand how our freedom is compatible with the realm of causality we belong to (Rel 135n.). We cannot say where or how the divine intervenes in spatio-temporal reality, though we can certainly hope and believe it does. We may even have to entertain a belief in grace if we are rationally to resume what has previously been so ill fated a task. But we cannot achieve knowledge about how grace works (that lies beyond our cognitive capacities). This and other related subjects belong to the *mysteries* of faith (Rel 129–38). We are best advised not to spend time worrying about such matters. All this becomes idle speculation if it does not manifest itself in upright willing.

KIERKEGAARD: Then you are saying we are rationally permitted to regard the divine as immanent in moral reasoning?[6] You further seem to hold that our sense of unbending obligation means (on the principle of *ought implies can)* that we possess the ability to renew our willing,[7] and that this sense of obligation, along with the very reanimation of our moral efforts it induces, may be taken as evidence of God's gracious intervention on our behalf?

KANT: Exactly. Our willing and grace are one and the same, depending on how they are conceived. No priority can be placed

[6] "Inward divine revelation is God's revelation to us through our own reason" (LPT 160).

[7] "A change of heart such as this must be possible because duty requires it" (Rel 67).

on them in time or logic. I might add that anyone familiar with my thinking would see this as a basic feature of my philosophy. For example, as early as the *Foundations of the Metaphysics of Morals* I try to show that our sense of freedom (given to us through the imperative of the moral law) is evidence of our "citizenship" in an intelligible realm (GR Part III). I never deny noumena or their presence in spatio-temporal realities, but I argue that our only insight into them is through the lens of our "natural" moral experience. The same is true of grace.

KIERKEGAARD: Then in a sense you are a "pure mystic" of reason, like one of those people you have a young friend describe in an appendix to *The Conflict of the Faculties?* (CF 70–75).

KANT: You are indeed a careful reader of my work, Kierkegaard!

KIERKEGAARD: Perhaps too careful. For now I must tell you that though I well understand your arguments in the *Religion* concerning grace, I must disagree with your conclusions about revealed faith and the historical savior. In fact, you might say much of my work is a development of this disagreement.

KANT: I am fascinated to hear you say that. Please explain.

KIERKEGAARD: Gladly. But before getting to specifics, let me see whether you agree that your argument in the *Religion* and *Conflict* has two prongs. One maintains that the concept of a historical savior contains ideas that are morally repugnant. The other maintains that such a historically based faith is not really needed.

KANT: I would like to see how you flesh out this broad categorization of my argument, but it seems correct.

KIERKEGAARD: Then let's take the first prong: that the concept of a historical savior is morally unacceptable. In the *Religion* you develop your understanding of the Son of God as an archetype in our reason of the ideal of a humanity well pleasing to God, but not as a historical person (Rel 55). Indeed, you deny that a living individual, however righteous, can ever be known to be anything more than a naturally begotten human being (Rel 57–59). This. I take it, is another implication of your denial of our knowledge of noumenal reality.[8]

KANT: Quite right.

KIERKEGAARD: But then you go on to say that if we were to elevate even a holy and righteous person to the status of a God-man, this could actually hinder our ability to adopt that person as a model for imitation because it would place him beyond all our normal human frailties and burdens (Rel 57).

KANT: Exactly. The effort to elevate a holy man in this way really defeats itself by rendering him utterly inapplicable to us. How can we learn to overcome bitter adversity, temptation, and fear if our model possesses a holy will and divine assurance of his own redemption through suffering and death? Although I didn't say this in the *Religion,* it seems to me that this impulse to elevate a man to the status of God can become an excuse for moral sloth and turpitude. Since we are merely weak and imperfect creatures, why should we aspire to moral perfection if only a God-man can accomplish it?

KIERKEGAARD: These are powerful points, but I believe they are fundamentally mistaken.

KANT: Why?

KIERKEGAARD: First, because you misconceive the nature of our savior, Jesus Christ. You ignore the teaching that he is "fully God and fully Man." In his human nature, he entirely enters into our trials and temptations, and he never draws on his divinity in the struggle toward goodness. Have you forgotten his prayers in Gethsemane? His agony of doubt and abandonment on the cross?

[8] This conception parallels Kierkegaard's denial that the "contemporary" is in any way privileged. *See* CUP1 97n.; PF 63; PC 62–66. The similarity between Kant and Kierkegaard on this matter is noted by Gordon E. Michalson, Jr., in his *Lessing's "Ugly Ditch": A Study of Theology and History* (University Park: Pennsylvania State University Press, 1985) 17.

Yet despite this, he never succumbs to sin, never relinquishes his holy mission. Christ for us is fully a model, never beyond what lies within our reach as human beings. And yet he is, in a sense, a negative model, as well.

KANT: A negative model? Do you mean that we should not imitate him?

KIERKEGAARD: No, just the opposite: we should and must imitate him, but we do not. Hence, Christ highlights our sin and deepens our remorse over our culpability for the abandonment of God's holy ways.

KANT: How intriguing. I confess I never saw things this way. But it is certainly a morally commendable idea.

KIERKEGAARD: More than commendable. I would say it is necessary for any moral life that seeks completion. Without this demanding, holy example to sharpen the requirement, we succumb to excuses and sink into the very moral lassitude you denounce. This is why I maintain throughout my work that a Christian ethic is truer to morality than any merely autonomously conceived moral law (PAP X2 A 396 n.d., 1850 [JP1-188]).

KANT: But you are not saying that we can dispense with autonomous reason?

KIERKEGAARD: No, moral reason interpenetrates our religious concepts. It leads us to an awareness of the inadequacy of our unaided efforts. But moral reason alone easily slips away from the requirement and provides opportunities to soften our self-judgment. Your own work provides many examples of how rational ethics has evaded these problems. The Stoic misuse of the concept of self-contentment to minimize the full challenge of moral commitment is one example. Your own evasion of the full implications of your discovery of the depths of sin is another.

KANT: My own evasion of the implications of sin?

KIERKEGAARD: Yes. Although you do an excellent job in the *Religion* and *Conflict* of establishing the depths of our defection from the moral law, your own solution to this problem—moral rededication without prior confession of God's saving act in Christ—is inadequate to the problem you develop. This returns me to the other prong of your argument against a historical savior: the claim that such an act of salvation is unnecessary for us. I believe just the opposite is true: that we cannot effect our full moral conversion *without* belief in such a saving act.

KANT: What is your argument for this?

KIERKEGAARD: In fact, you are the one who supplied me with the argument, both in the *Religion* and *Conflict*. You know that I pored over both these works and even copied out passages and bits of humor from them into my journals.[9] Your bold moral denunciation in both texts of Abraham's willingness to obey God's command in Genesis 22 was a major stimulus to my presentation of the ethical in *Fear and Trembling*.

KANT: I'm aware of your attention to both my works. But how do they contribute to your criticism of me?

KIERKEGAARD: Reading and rereading these two books, I came to see them as a progressive involvement with the question: How are we to recover from the depths of moral self-judgment that our reason discloses to us? In the *Religion*, your answer to this revolves around your rationalized and moralized conception of divine grace as evidenced in our own sense of unbending obligation to moral reform and renewal. Grace enters to provide a rational grounding for what might otherwise appear irrational.

KANT: Quite right. In the *Religion* divine grace provides an answer to the question "How is moral reform logically possible?" just as a priori knowledge enters in the first *Critique* as an answer to the question "How is experience logically possible?" and as freedom enters in the second *Critique* as an answer to the question "How is morality logically possible?"

---

[9] For references to these passages, see Green, *Kierkegaard and Kant*, 9–31.

KIERKEGAARD: But you will admit that in both the *Religion* and *Conflict* you equivocate about whether we must openly confess to ourselves our full reliance on God's grace?

KANT: What do you mean?

KIERKEGAARD: On the one hand, you develop the depths of our sin. You point out that we must judge ourselves guilty of infinite sin. You acknowledge that if we were to stand before a righteous judge, we would pass judgment on ourselves with the greatest severity (Rel 64); you even concede that, but for our inability to see beyond the limits of this life, we would reasonably go from mere comfortlessness about our eternal state to "wild despair" (Rel 65).

KANT: That is all true. I also say, as you acknowledge, that we find the confidence we need to carry on in our reformed moral disposition, a disposition that is grace enough for us.

KIERKEGAARD: But what allows us to regard this as anything more than self-deception? Recall your own remark that "man is never more easily deceived than in what promotes his good opinion of himself" (Rel 62). Isn't this reliance on our disposition an evasion of the depths of the problem?

KANT: I acknowledge that our sense of requirement and empowerment may occasionally be self-deluding. But what is the alternative? To acknowledge ourselves as fatally mired in sin? To give up and abandon reform? What is the profit in that? What is wrong with taking our indwelling sense of obligation as a sign of grace and proceeding from that?

KIERKEGAARD: Two things. First, it is an invitation to those who feel any moral urgings at all to ignore the significance of their defections from the moral requirement. Second, it provides license to repeat the errors of the past and even to seek new occasions for self-assertion under the guise of moral renewal.

KANT: Granted that is a risk: But, again, what is the alternative? Total, wild despair? Self-indulgent wallowing in our own moral incapacity and wickedness? Let me take a leaf from your book, *The Sickness unto Death,* and ask whether what you are counseling doesn't amount to "despair over forgiveness"?

KIERKEGAARD: I appreciate your attention to my work. But I think you miss my point. My aim in *The Sickness unto Death* was to drive us to God's forgiveness in Jesus Christ. It is to avoid this that the prideful, demonic personality despairs over forgiveness and wallows in condemnation. The willingness to accept forgiveness in and through God's atonement for us in Christ is the alternative both to despair and to shallow, autonomous moral self-renewal. Furthermore, it is an alternative that doesn't enervate but energizes the moral life.

KANT: How is that?

KIERKEGAARD: First, it forces us to strip away all our self-deception and naïve confidences and greatly intensifies our sense of the requirement. As I showed in the *Philosophical Fragments,* there is all the difference between a teacher who merely reminds us of what we already know and one who shows us how deeply in error we are (PF 9–22). Second, the fact that God has actually entered time, suffered, and died on our behalf provides us a real basis for the confidence that we can be and have been forgiven. In saying this I am again drawing on your writings.

KANT: How so?

KIERKEGAARD: I employ your point in the first *Critique* that there is a significant difference between *logical* possibility and *reality* (KRV 479). Many things are logically possible. They belong to what I call the sphere of "ideality" or "concept existence" (PF 39–42). But only some things are actually given to us in experience, as you say, and, hence, are really possible. A God-man and our enduring moral redemption are both logically possible. Your work shows that. Neither can be refuted, unless we succumb to a dogmatic empiricism that denies that noumena can be expressed in time. But it is one thing to say that something is logically possible and quite another to say it has come to pass. My point is that the depth of our valid moral despair requires *real,* not logical, possibility.

KANT: You make a powerful case.

KIERKEGAARD: Actually, once again it is your case. You develop it in *Conflict.*

KANT: Do I?

KIERKEGAARD: Yes, in an oblique way. Do you recall the imaginary dialogue you present there in the form of a series of objections on the part of a defender of revealed faith to "a rational interpretation" of the Bible?

KANT: I would have to search my memory.

KIERKEGAARD: Let me help you. I actually have a copy of the text here on this wonderful laptop computer I purchased.

Kierkegaard withdraws a portable computer from his traveling bag and places it on the table before them. He types several keystrokes, and smiles with satisfaction.

KIERKEGAARD: These computers are wonderful things, aren't they? You and I owe our very existence to them. But here is the dialogue, just as I remembered it. One of the objections you consider is stated as follows:

> To believe that God, by an act of kindness, will in some unknown way fill what is lacking to our justification is to assume gratuitously a cause that will satisfy the need we feel (it is to commit a *petitio principii*): for when we expect something by the grace of a superior, we cannot assume that we must get it as a matter of course: we can expect it only if it was actually promised to us.... So it seems that we can hope for that supplement and assume that we shall get it only in so far as it has been actually pledged through divine *revelation.* (CF 83)

This seems to me to be a powerful argument.

KANT: What is my rationalist reply?

KIERKEGAARD: What you say is the following:

> A direct revelation from God embodied in the comforting statement "Your sins are forgiven you" would be a supersensible experience, and this is impossible. But it is also unnecessary with regard to what (like religion) is based on moral principles of reason and is therefore certain a priori, at least for practical purposes. There is no other way we can conceive the decrees of a holy and benevolent law-giver with regard to frail creatures who are yet striving with all their might to fulfill whatever they recognize as their duty; and if, without the aid of a definite, empirically given promise, we have a rational faith and trust in His help, we show better evidence of a pure moral attitude and so of our receptivity to the manifestation of grace we hope for than we could by empirical belief. (CF 83)

Now, my dear Kant, I hope you will agree with me that this reply, so central to our disagreement over the need for faith in a historical savior, is a pastiche of misleading and unrelated ideas.

KANT: Isn't that extreme? What is wrong with these remarks?

KIERKEGAARD: Take the first part of the reply that a direct revelation from God regarding forgiveness "is impossible." Certainly this misstates the matter. It is by no means impossible that a statement or revelation to the effect that we are forgiven should come from God. Only that we should *know* such a statement to be from God is impossible. Your whole philosophy rejects an empiricist dogmatism that rules out the possibility of noumena amidst phenomena.

KANT: Of course you are right. What I meant to say is that it is impossible for us to make an assertion with any claim to "knowledge" that a particular communication (whether through words or events) actually proceeds from God.

KIERKEGAARD: On this matter there is no disagreement between us. In *Training in Christianity* and elsewhere I emphatically deny that the contemporary believer who witnesses

Christ in his midst has any advantage over those of us, centuries later, who attest in faith to the meaning of that life. We are talking here about faith, not knowledge. For you, it is a rational faith based on moral concepts. For me, it is a moral faith graciously given by God. But surely it is misleading to say, as you seem to do, that such a revelation itself is impossible when what you mean is the far more modest observation that we cannot possess *knowledge* that a revelation is from God. I might add, by the way, that where God's forgiveness of us is concerned, we are not talking about a single oracular assertion (out of the mouth of a fanatic like Swedenborg) but about the entire record of Christ's holy life—and death.

KANT: You're right about all this. The rationalist's reply does not represent my own point of view here.

KIERKEGAARD: Your subsequent assertions are no more helpful. You say a direct revelation from God that we will receive his help is unnecessary because there is "no other way" we can conceive the decrees of a righteous law-giver with regard to "frail" but morally striving creatures like ourselves. But this merely restates your position. It is no reply to the objection that even when we have good reason to "expect something by the grace of a superior we cannot assume that we must get it as a matter of course." I think what the objector is trying to say here is that we would be morally presumptuous to interpret *our need* as somehow creating a requirement for God to meet it. Such presumptuousness is deeply contradictory to the humility that should characterize those who have come to recognize their sinfulness. But if we have no right to presume that God will aid us, we return to the question of how those who are mired in sin can gain the reasonable confidence they need to renew their striving. Revelation can provide this by offering some evidence (not certain evidence but enough to justify confidence) that God *has* committed himself to our redemption.

KANT: An interesting moral point. I like to think of myself as a master of practical rational arguments, but I am not accustomed to having moral reason turned against me as you do here. I confess that I can see nothing wrong in what you say.

KIERKEGAARD: Your final point in this reply strikes me as equally uncompelling.

KANT: Are you referring to my assertion that we show better receptivity to the manifestation of grace if we rely on faith and trust in God's help rather than on any definite, empirically given promise?

KIERKEGAARD: Yes. I'm aware of how central this assertion is to your practical philosophy. An analogue to it appears at the end of the second *Critique* when you say we are better off lacking knowledge of God's existence because this affords us the opportunity to develop a purer moral disposition unaffected by the certainty of reward or punishment (KPV 147). I take it you are trying to say the same thing here: that our moral sensibilities are sharpened by the absence of security regarding our redemption. In terms familiar to us both, you are saying that we must work out our salvation in "fear and trembling."

KANT: Again you impress me with your grasp of my writings.

KIERKEGAARD: But I must say that this is the one place where such uncertainty may be inappropriate. Recall that our morally committed person who has come to see the depths of his or her sin does not face a problem of false confidence and assurance, but just the opposite: a paralyzing state of self-condemnation in which every buoying reflection must be rejected as merely another expression of self-deception. Here, it seems to me, strong medicine is needed. And since nothing is morally or cognitively wrong with perceiving the Gospel record as a sign of God's love for us, we are fully warranted in making the events of Christ's holy life, death, and resurrection the conceptual starting point for our moral rebirth. Acceptance of these beliefs is merely another expression of your own insistence on the priority of practical reason and an extension of your willingness to accept noumenal beliefs that do not violate theoretical reason on grounds provided by compelling convictions of our practical reason.[10] In this case, practical need triumphs not only over the ordinary cautions of theoretical reason (such as not easily admitting belief in a God-man, though it cannot condemn such a belief), but also over moral reason and its reluctance to qualify moral autonomy in any way.

KANT: Your arguments are very compelling. If I hesitate to accept them with enthusiasm, this is because during my life I witnessed so many fanatical Christian believers who insisted on making this belief in saving grace the starting point, but who then engaged in "passive surrender" to grace (CF 75) and never thereafter demonstrated the effort of moral renewal it was supposed to lead to.

KIERKEGAARD: I share your concern. My own writings, as you may know, criticize those Christians who appeal to grace and then forget the "requirement." But it is one thing to object to such weakness and misuse of concepts and another thing to reject those concepts themselves. The reply to our tepid Christians is not to ignore the Gospel but to preach it in its full depth and rigor. That is what I tried to do during the last phase of my life in what is called my "attack on Christendom."

KANT: I am familiar with your efforts. They showed great courage and perseverance.

KIERKEGAARD: Thank you, though I know I was also motivated by more than a little bitterness. I would like to accept your compliment, but I realize that "man is never more easily deceived than in what promotes his good opinion of himself" (Rel 62).

[10] See Allen W. Wood, "Kant's Deism," in *Kant's Philosophy of Religion Reconsidered*, ed. Philip J. Rossi and Michael Wreen (Bloomington: Indiana University Press, 1991) 17–18.

KANT: A question. I am now in a better position to understand the outlines of your sustained philosophical defense of revealed Christian faith. I am even persuaded that the concepts you develop may be acceptable to our morally legislative practical reason. But you are not saying that these beliefs are *required* by reason, are you? If so, I would have two problems with such a position. First, as I indicate in *Conflict*, I am opposed to the imposition of dogmas. These dogmas, so dear to ecclesiastical faith, involve matters we can never settle with certainty and are an invitation to needless strife and conflict. Second, to go beyond those concepts directly given to us by our practical reason and require beliefs based on revelation is to exclude from moral salvation whole sectors of humanity not privy to such revelation. In other words, these revealed beliefs lack the universality appropriate to fundamental moral concepts (Rel 95; CF 77).

KIERKEGAARD: I have given little thought to the question you raise. It is the question, really, of whether Christian faith itself, especially a confession of belief in Christ, is required for our moral salvation. Let me begin to answer your question by observing that it is really two questions.

KANT: How so?

KIERKEGAARD: First, it is the question of whether we need some real evidence in experience in order to build a commitment to moral renewal. That is, must our moral rededication be accompanied by certain theoretically or empirically warranted beliefs to ground the confidence that we are anything other than morally doomed? To this question, I must give a strong affirmative answer. Everything both you and I have said about the totality of moral despair leads me to believe that without some warrant beyond the voice of conscience, we must end in the paralysis of self-condemning moral despair. I might add here that your own writings repeatedly affirm the need for some rationally based confidence—a reasonable hope (Rel 46; CF 75–77)—in at least the possibility of assistance in our moral renewal.

KANT: I grant that. Although I try to dissuade my readers from dwelling on this matter (because I believe it has too often distracted people from the practical task at hand), I do observe that reason cannot put such "speculative" questions aside lest it be accused of "being wholly unable to reconcile with divine justice man's hope of absolution from his guilt" (Rel 70). I take it you would argue that such "speculative" matters, bound up with the question of whether we have been and can be saved. are far more important than I am inclined to admit.

KIERKEGAARD: Yes, though I see myself as developing, rather than contradicting, your position.

KANT: Fair enough.

KIERKEGAARD: But to return to your earlier question of whether we are rationally required to believe in Christ, a second question is implicit in this. This is the question of whether, in addition to having a well-founded belief to base our moral rededication on, this belief must focus on God's activity in Jesus Christ. In other words, to what extent is this one historical faith requisite for salvation?

KANT: I see that question. It is one that preoccupies me. Let me repeat my specific apprehensions. How can we make human beings' moral redemption depend on a faith known only by some, with all the implications this has for dogmatic tyranny, exclusion, and oppression?

KIERKEGAARD: I share your concerns. Let me say, as odd as it may seem, that this is not a question I really addressed in my writings. Remember who my intellectual opponents were: First, the cultured despisers of Christianity, especially the Hegelians, who denied that we even needed the faith of the Gospel for our ethico-religious life. These were the smug, unthinking, and very distant heirs to aspects of your philosophy, convinced that we are the best of men living in the best of times. Against them I used the moral rigor I learned from you—something wholly lacking among the Hegelians—to intensify the moral demand and to develop the importance and value of historical Christian faith. But at no time was I called on philosophically to defend the rational necessity of Christian belief in general. My second set of opponents was the lax residents of "Christendom" who hardly needed to be told that Christian faith is necessary. What had to be assaulted was their confidence that they already had such faith by virtue of their possession of a baptismal certificate or the fact of their birth in a "Christian" nation. Hence, a defense of Christian faith against other religious or philosophical positions was never really my concern.

KANT: I understand that. But how, then, do you answer my question? Do you believe a faith in the promise of God's forgiveness through Christ is rationally required for our salvation?

KIERKEGAARD: This is a difficult question to answer. Frankly, I'm not sure. On the one hand, like you, I think, I would probably say that what matters is not our words or even beliefs but our ethico-religious passion. A person who comes to despair over the rigor of the ethico-religious demand and who then avoids despair's complex evasions and accepts the requirement fully is on the way to salvation. On the other hand, I personally find it difficult to see how one can come to know either the requirement or the possibility of meeting it apart from Christ.

KANT: Would you say, then, to do so is impossible? Are there perhaps other philosophical or religious paths that might take us to these insights? Perhaps they exist in religious traditions that you and I know little about? If you have read my *Religion* carefully, you will note that I frequently suggest that basic biblical insights have their counterpoints in other religious traditions (Rel 15, 68n., 131n.). Are the concepts requisite for salvation from a deepened awareness of our moral failure also to be found elsewhere? This is important because it would address one aspect of my insistence that the religious-moral conceptions we require must be universal, though they might take different forms for different peoples.

KIERKEGAARD: I suppose that is possible. But I repeat, this was never my concern, never a part of the challenge I faced. My life task was neither to convert nor even to consider the pagans but to remind those living in Christendom and claiming to be Christians of the meaning—and the demands—of their faith.

KANT: I appreciate that. In this sense, at a different moment in the development of Christianity, we were not so far apart. For my concern in fighting against historically based ecclesiastical faith was to combat the ecclesiastical faith whose dogmatism had led to moral laxity, conflict, and violence.

The waiter approaches to say that both men's flights have come up on the departure screen of the Club's computer and that they had better start for their boarding gates.

KANT: Well, my dear Kierkegaard, we must break off in mid-course. Perhaps we can turn to some of these larger questions another time. This has truly been a pleasure. I genuinely hope this is only the beginning of a sustained conversation between us. I would like to invite you to Königsberg for a more formal discussion of these matters with others in attendance, but, as I'm sure you know, my natal city, though renamed, is still being reconstructed and lacks the resources for an adequate scholarly conference.

KIERKEGAARD: Perhaps some time in Copenhagen? Or better yet, why not San Francisco, a charming city with good wine and a far more pleasant climate than either of our Baltic homelands!

KANT: Agreed! Auf Wiesdersehen.

KIERKEGAARD: Auf Wiesdersehen. Farvel!

5

Kant: A Debt both Obscure and Enormous

Kierkegaard's debt to Kant is at once obscure and enormous. There are only seventeen explicit references to Kant in the works published by Kierkegaard during his lifetime. This compares with more than three hundred references to Hegel, fifty-five to Aristotle and more than ninety to Kierkegaard's Danish philosophical contemporary H.L. Martensen. But despite this paucity of overt references, Kant is a major presence in some of Kierkegaard's most important pseudonymous books, including *Either/Or, Fear and Trembling, Philosophical Fragments,* and the *Postscript*. Kant's thought also figures predominantly in two of Kierkegaard's most important religious discourses, *Works of Love* and *Purity of Heart*. In what follows, I will suggest that although Kierkegaard clearly rejected some aspects of Kant's rationalist philosophy, he more often used it as the philosophical platform on which to construct his own unique religious authorship.

Even though Kant was going out of vogue among Danish intellectuals during Kierkegaard's university years, there is no doubt that Kierkegaard paid close attention to the German philosopher's writings. Fifteen of the ninety-nine questions on Kierkegaard's 1840 oral examination for the theological degree (*Attestats*) were directed at ethics and philosophy of religion, with Kant named explicitly in two of them and his work alluded to in most of the others.[1] The *Auktionsprotokol*, or auction catalogue of

[1] This record is published in *Breve og Aktstykker vedrørende Søren Kierkegaard*, ved Niels Thulstrup (Copenhagen: Munksgaard, 1953) I, 9–12.

Kierkegaard's library at the time of his death, indicates that Kierkegaard's library included the *Critique of Pure Reason* and the *Critique of Judgment* as well as an edition of Kant's minor treatises (*vermischte Schriften*) edited by Johann Heinrich Tieftrunk and published in 1799 in Halle, Germany. On the basis of many references, it is highly likely that he owned and perused Kant's *Religion within the Limits of Reason Alone* as well.

There are many intersecting signs that Kierkegaard carefully read most of Kant's principal writings, as well as some of his lesser-known essays.[2] Some of these works are mentioned by name or clearly alluded to in Kierkegaard's journals and papers. Perhaps the most important of them for our understanding of the impact of Kant on Kierkegaard is Kant's 1798 work *The Conflict of the Faculties*. Other works by Kant make their influence known by Kierkegaard's use of or reliance on important terms and concepts present in them. Thus, although Kierkegaard never explicitly references Kant's *Critique of Pure Reason*, whole sections of the *Philosophical Fragments* evidence his knowledge of this work's critique of the ontological argument for God's existence, right down to Kierkegaard's appeal to Kant's humorous illustrations. Similarly, Kant's *Religion within the Limits of Reason Alone* makes its presence known in Kierkegaard's thinking through his occasional reference to the *Religion*'s concept of "radical evil."

In my book *Kierkegaard and Kant: The Hidden Debt* I try to systematically develop all the points of contact and deeper engagements between Kierkegaard and Kant. I do not intend to retrace all those matters here. Instead, I want to use the arguments in several of Kant's and Kierkegaard's key writings to illustrate the extraordinary degree of Kierkegaard's indebtedness to Kant both as an intellectual adversary and as a teacher.

---

[2] For a fuller discussion of Kierkegaard's readings in Kant, see my *Kierkegaard and Kant: The Hidden Debt* (Albany: State University of New York Press, 1992) 9–31.

The Conflict of the Faculties is a good place to begin. Published near the end of Kant's life, and five years after the *Religion,* it represents Kant's most mature treatment of the relationship between revealed religion and rational philosophy. The issue is treated in the first (and longest) part of the book entitled "The Conflict of the Philosophy Faculty with the Theology Faculty."[3] Kant's purpose in this treatise is to provide some broad principles for adjudicating recurrent disputes about the meaning of revealed (biblical) religion that arise between "biblical theologians," on the one hand, and rationalist philosophers (like himself), on the other. These include such matters as how we are to understand the doctrines of the Trinity, the resurrection of Christ, or the teaching of predestination. Kant's solution to such conflicts is to insist on the right of philosophy to establish the rational limits of revealed teachings. "In matters of religion, reason is the highest interpreter of Scripture" (CF 71). It follows that when scriptural teachings appear to contradict moral reason, they must be "re-interpreted in the interests of practical reason" (CF 65).

Applying this methodological approach, the balance of the argument of *The Conflict of the Faculties* then turns to the development of two interrelated themes: the significance of divine grace and the role of belief in the historical bestowal of that grace in our moral salvation. Kant's view is clear: divine grace is a rationally acceptable concept, but it must never be understood in ways that impugn man's moral freedom and responsibility.

In *The Conflict,* Kant does not make the argument that a concept of grace is rationally permissible and perhaps even rationally required. That argument has been developed extensively in the predecessor work, *Religion within the Limits of Reason Alone,* to which I will turn shortly. But he does here

[3] The remainder of the CF, which looks at the conflicts between the philosophy faculty of the university and the faculties of law and medicine, is briefer and contains nothing of religious importance.

summarize his earlier conclusions: "If man's own deeds are not sufficient to justify him before his conscience (as it judges him strictly), reason is entitled to adopt on faith a pure supernatural supplement to fill what is lacking to his justification…" (CF 75). Nevertheless, this concession to divine activity in the economy of our moral salvation must never be interpreted so as to license an abandonment of man's obligation to constant moral striving:

> Actions must be represented as issuing from man's own use of his moral powers, not as an effect [resulting] from the influence of an external, higher cause by whose activity man is passively healed. The interpretation of scriptural texts which, taken literally, seem to contain the latter view must therefore be deliberately directed toward making them consistent with the former view. (CF 73–75)

In the ensuing development of this theme, Kant takes seriously a host of views associated with various Protestant sects in his day (Pietists and Moravians) that wrestle with the question of how we can think of divine grace as being integrated with the experience of moral self-renewal. Ultimately, Kant dismisses these views as so much useless speculation on matters that are essentially beyond our ken (because they deal with supersensible realities that transcend the bounds of human knowledge). He also judges them to be irrelevant since all that needs to concern us is our own renewed moral striving. The presence of such striving, Kant argues, is both our evidence of divine grace and our most appropriate response to it. We can put aside the speculative question of *how* such striving is possible for us.

Having laid out his own views on the agency of grace, Kant observes that "reason has its own objections" (CF 81) to his position. One of these goes to the heart of his claim that as rational beings we are warranted in assuming divine grace as a needed supplement to our imperfect moral efforts.

> *Objection*: To believe that God, by an act of kindness, will in some unknown way fill what is lacking to our justification is to assume gratuitously a cause that will satisfy the need we feel (it is to commit a *petitio principii*); for when we expect something by the grace of a superior, we cannot assume that we must get it as a matter of course; we can expect it only if it was actually promised to us,... as in a formal contract. So it seems that we can hope for that supplement and assume that we shall get it only in so far as it has actually been pledged through divine *revelation*, not as a stroke of luck. (CF 83)

Kant immediately replies to this objection. The reply embraces several familiar Kantian themes. First, that as creatures whose knowledge is always linked to sensory experience, we are cognitively unequipped to recognize a divine bestowal of grace: "A direct revelation from God embodied in the comforting statement 'Your sins are forgiven you' would be a supersensible experience, and this is impossible" (CF 83). Second, he insists that in a matter of practical reason like this, the *knowledge* that we have received such forgiveness is both unnecessary and undesirable. As frail creatures striving with all our might to fulfill what we recognize as our duty "there is no other way we can conceive the decrees of a holy and benevolent law-giver" than to assume that we are eligible for his forgiveness. Furthermore, "if, without the aid of a definite, empirically given promise, we have a rational faith and trust in His help, we show better evidence of a pure moral attitude and so of receptivity to the manifestation of grace we hope for than we could by empirical belief" (CF 83).

I must intervene in the middle of Kant's argument here to observe that the themes he is exploring and issues he is engaging are pervasive in Kierkegaard's writings and are absolutely central to Kierkegaard's concerns in two of his most important pseudonymous works, *Philosophical Fragments* and *Concluding Unscientific Postscript*. Both works deal with the question of whether human beings can complete the task of their moral salvation on their own through the use of rational concepts, or whether they must transcend these concepts in a leap of faith to admit the possibility of God's historically bestowed grace. Kant's insistence in *The Conflict* that although we may presume a role for grace in our salvation, we have all that we need in our own rational conceptions of God and morality to surmount moral failure and sin places him squarely in the camp of rationalist philosophy and ethics. Kant thus epitomizes the "Socratic" viewpoint that Kierkegaard presents and transcends in the *Fragments*.

The relevance of Kant's discussion here to Kierkegaard is signaled by a remark by Kant somewhat earlier in the treatise. Distinguishing *pure religious faith,* which is based on "mere reason," from *ecclesiastical faith,* which is based entirely on revealed statutes, Kant dismisses the latter outright:

> The biblical theologian says: "Search the Scriptures, where you think you find eternal life." But since our moral improvement is the sole condition of eternal life, the only way we can find eternal life in any Scripture whatsoever is by putting it there. For the concepts and principles required for eternal life cannot really be learned from anyone else: *the teacher's exposition is only the occasion* [Veranlassung] *for him to develop them out of his own reason.*[4] (CF 63)

In the *Fragments,* of course, Johannes Climacus offers as the hallmark of the Socratic position that the teacher is nothing more than an occasion (he uses the cognate Danish word, *Anledning*) for the student's own recollection of what is eternally true. Climacus uses this term more than forty times there in connection with the Socratic standpoint, and explicitly attributes it to Socrates, whom

[4] The italics in the translated passage are my own. The German for the latter part of this passage reads: "... *weil die dazu erforderlichen Begriffe und Gründsatze eigentlich nicht von irgend einem andern gelernt, sondern nur bei Veranlassung eines Vortrages aus der eigenen Vernunft des Lehrers entwickelt werden müssen*" (CF 62).

he presents as introducing it in the *Meno*.[5] But while a pedagogical relationship is assumed in the famous portion of the *Meno* where the young slave is made to "recollect" the truths of geometry, the term "occasion" is never used. Similarly, this conjunction of terms is not found elsewhere in Plato's writings or in any of the rationalistic philosophers preceding Kierkegaard—except for Kant. Can it be merely coincidental that Kierkegaard chooses the very same term to describe the rationalist view of saving knowledge that Kant uses to contrast his own rationalist position with a historically and scripturally based revelational position?

The importance for Kierkegaard of Kant's discussion in *The Conflict* becomes even more apparent when Kant applies his general dismissal of supernatural agency to the issue of the need for historical faith. This theme, which occupies the remainder of the discussion of religion in *The Conflict,* follows from the discussion of grace. Since we have inherent rational concepts that enable us to rededicate our moral lives, what good is served by scriptural testimonies to God's intervention in history? Kant's answer, essentially, is "none at all." "[W]e must regard the credentials of the Bible as drawn from the pure spring of universal rational religion dwelling in every ordinary man; and it is this very simplicity that accounts for the Bible's extremely widespread and powerful influence on the hearts of the people" (CF 115–117). The Bible itself is thus merely "the vehicle of religion," whose spirit is its universally accessible rational moral content. The letter of its decrees "do not belong to what is essential in it (*principale*) but only to what is associated with this (*accessorium*) [additional element]." It follows from this that "No historical account can verify the divine origin of such a writing. The proof can be derived only from its tested power to establish religion in the human heart" (CF 117). Kant closes this discussion with a direct

[5] Kierkegaard (PF 9) cites the passage as "80, near the end." The explicit discussion of this matter actually extends from 82–86.

attack on the appeal to the authority of the revealed text itself: "[I]t is superstition [*Aberglaube*] to hold that historical belief is a duty and essential to salvation" (CF 119).

It is not hard to see that Kant here is forcefully enunciating the very position that Kierkegaard attributes to Socrates and challenges in the *Fragments* and *Postscript*. A discussion appearing only a page before the above remarks furnishes a powerful further reminder that it is Kant's articulation of this position (not that of Socrates or anyone else) that is on Kierkegaard's mind. Here Kant voices the most bitter criticism of Abraham found anywhere in the philosophical tradition up to that time. The occasion for this criticism is Kant's acknowledgement that a law book of God's statutory, revealed will harmonizing perfectly with morally practical reason would be an effective organ for guiding men and citizens to their temporal and eternal well-being, "if only it could be accredited as the word of God and its authenticity could be proved by documents." But, Kant quickly adds, "there are many difficulties in the way of validating" such a law book. The first difficulty rests on Kant's denial that human beings can ever "know" supersensible truths: "if God should really speak to man, man could still never know that it was God speaking. It is quite impossible for man to apprehend the infinite by his senses, distinguish it from sensible beings, and *recognize* it as such" (CF 115). The second difficulty is that the Bible appears to contain teachings that run directly contrary to human moral reason. In such cases, says Kant, "man can be sure that the voice he hears is *not* God's" and "he must consider it an illusion" (CF 115). A footnote here takes us to directly to the text that Kant has in mind:

> We can use, as an example, the myth of the sacrifice that Abraham was going to make by butchering and burning his only son at God's command (the poor, child, without knowing it, even brought the wood for the fire). Abraham should have replied to this supposedly divine voice: "That I ought not to kill my good

son is quite certain. But that you, this apparition, are God—of that I am not certain, and never can be, not even if this voice rings down to me from (visible) heaven.[6] (CF 115)

Kant's criticism of the Abraham "myth" (and with it of Abraham himself, the biblical narratives, and, perhaps the God portrayed in them) is essentially moral in nature. It is not hard to imagine its impact on Kierkegaard, whose pseudonym, Johannes de Silentio, in *Fear and Trembling* will take up the challenge of defending Abraham, God, and biblical revelation. But the significance of Kant's criticism of the Genesis 22 narrative for Kierkegaard goes far beyond the narrow moral issues at stake. For as Kant well knows, and makes clear in another footnote just several pages further on in *The Conflict*, the Abraham story has fundamental significance for Christian thought. There Kant alludes to an idea he reports as prevalent among Jesus' immediate Jewish followers that Greeks and Romans could also be regarded as admitted into the new Jewish covenant if they believed in Abraham's sacrifice of his only son and interpreted it "as the symbol of the world-savior's own sacrifice" (CF 121). Kant immediately rejects this idea, however, adding that "if a church commands us to believe such a dogma, as necessary for salvation," and we obey out of fear, our belief is superstition" (CF 121).

This equation of the father-son pair, Abraham and Isaac, with God and Christ is a staple of Christian theology, having its scriptural beginning in New Testament texts.[7] This tells us that the issue at stake for Kant here is not simply Abraham's ethics: it is the relevance of the historical reports found in scripture to our

---

[6] There is a typographical error in Mary Gregor's English text "not even is this voice" that I have corrected here.

[7] Hebrews 11:17–19. For a fuller account of this tradition, see David Lerch, *Isaaks Opferung christlich gedeutet* (Tübingen: J.C.B. Mohr, 1950). Kierkegaard's familiarity with this tradition is evidenced by a journal entry for 1839—PAP II A 569 13 September 1839 [JPI—298].

salvation. This impression is reinforced when we note that Kant also treats Genesis 22 in the *Religion* in the context of two discussions of the reliability of historical human reports in matters of faith and ethics. One discussion denies historical reports of miracles any place in the formation of faith and concludes by saying that it is essential that "in the use of these historical accounts, we do not make it a tenet of religion that the knowing, believing, and professing of them are themselves means whereby we can render ourselves well-pleasing to God" (Rel 80). Genesis 22 is then introduced as a "theistic miracle" that "flatly contradicts morality" and as such, "cannot, despite all appearances, be of God" (Rel 82). Further on in the text, Kant rejects the position of "an inquisitor" who appeals to a "supernaturally revealed Divine will" to ground his persecution of heretics. Such a person, Kant says, certainly acts in a way that moral reason can never countenance, "but that God has ever uttered this terrible injunction can only be asserted on the basis of historical documents and is never apodictically certain" (Rel 175). Even if God had directly spoken to the inquisitor, as he did to Abraham in telling him "to slaughter his own son like a sheep," there is still the chance "that in this instance a mistake has prevailed" (Rel 175).

We can see, therefore, that for Kant the import of the Genesis 22 narrative goes beyond the moral question of the relative priority of religious commands over obedience to moral reason. Rather, that issue contributes to his general rejection of the authority of historically based revelational claims. Precisely because we possess everything we need in ourselves to make judgments concerning our "temporal and eternal well-being," we have no essential need of scriptural guidance or promises, and these can even mislead us.

It is not hard to see that the *Fragments* and large portions of the *Postscript* dealing with historical faith can be viewed as rejoinders to Kant. If we also keep in mind the possibility, as I have argued for elsewhere,[8] that *Fear and Trembling* deals less with the theme of ethics than of soteriology, that it uses the *figure* of Abraham to defend a Pauline-Lutheran claim that God can suspend ethical judgment by entering history to forgive human sin, then three of Kierkegaard's major pseudonymous works can be seen as being provoked by Kant's discussion in *The Conflict of the Faculties*.

Anyone who has followed my argument to this point and found it persuasive might conclude that Kierkegaard was stimulated to embark on his defense of historical Christian faith because of his ardent opposition to the position outlined by Kant in *The Conflict of the Faculties*. But that conclusion would miss the most important and positive aspects of Kierkegaard's relationship to Kant. For while Kierkegaard strenuously disagreed with Kant's views on the importance of historical faith, that disagreement was largely built on Kantian foundations. Kierkegaard did not just disagree with Kant, he actively drew on Kant's writings on epistemology and ethics to construct his own unique defense of historical Christian faith.

I could trace multiple lines of borrowing here, but two in particular stand out: one ethical, the other epistemological. The ethical debt derives from Kant's development of the concept of "radical evil" in the *Religion*. To understand the importance of this concept, we must see that with it, the arch-rationalist enlightenment philosopher, whose constant emphasis is on human moral autonomy and human moral initiative, comes to acknowledge the inevitable and utter failure of even the best and most concerted moral efforts.

So striking is Kant's rejection here of man's ability to achieve the goal of moral righteousness that some commentators have

---

[8] "'Developing' *Fear and Trembling*," (reprinted here as chapter six); "Enough Is Enough! *Fear and Trembling* is *Not* about Ethics."

seen it as either Kant's cowardly concession in his old age to pressures from religious orthodoxy or an incoherent afterthought to his rationalist philosophy inspired by Kant's Christian upbringing.[9] Neither interpretation is correct. The *Religion*'s concept of radical evil is well prepared in the second part (or "Dialectic") of the *Critique of Practical Reason*. There, after having constructed his ethical position in the *Foundations of the Metaphysics of Morals* and the first part (or "Analytic") of this second *Critique*, Kant opened the way to his whole philosophy of religion by developing the understanding that no one is rationally required to adopt the standpoint of judgment represented by the supreme principle of morality, the categorical imperative. That imperative asks us to subordinate all our volitions and possible actions to a rule of universal acceptability as law. In choosing to act, we must put aside our own inclinations and ask whether the policy implicit in our proposed action (maxim) is something we could raise to the level of a law for all rational persons and is the kind of policy all such persons might rationally accept.[10]

This standpoint of judgment can obviously run counter to my real interests. For example, it may require me to forego making and then breaking a promise if, in doing so I can advance my interests. In such instances, the question arises, why should I adopt this rational standpoint? In a powerful series of arguments in passages of the second *Critique*, Kant recognizes that all previous arguments defending the moral point of view either

[9] It is a chestnut of Kant lore to view the *Religion* as Kant's effort to please his aging manservant Lampe. For the development of the view that Kant's concepts of grace reflect his unassimilated Pietist background, see Gordon E. Michalson, Jr., *Fallen Freedom: Kant on Radical Evil and Moral Regeneration* (Cambridge: Cambridge University Press, 1990).

[10] For a fuller development of Kant's conception of the categorical imperative, see my article "The First Formulation of the Categorical Imperative as Literally a 'Legislative' Metaphor," (reprinted here as chapter one).

reduce it to some misleading form of enlightened self-interest or beg the question by presuming the very standpoint they seek to commend.[11] This leads Kant to the conclusion that we possess a radical degree of freedom with respect to morality. Not only are we free from outer coercion in this area—this follows from our very freedom to make moral choices—but we are also free from the *rational* necessity to be moral.

In the second section ("Dialectic") of the second *Critique* Kant offers a complex argument for the role of religious faith in addressing this problem and facilitating human moral choice. Sometimes mistakenly called his "moral proof" of the existence of God, the argument is more properly thought of as an exploration of the conceptual presuppositions of rational moral commitment. Each of us as a human being, says Kant, is positioned between the unbending edicts of our rational moral conscience and our rational need to seek the fulfillment of our personal wishes in order to achieve happiness. Morality often imposes a stark choice between these two conflicting rational standpoints, and when this happens, it seems that we cannot make a choice in either direction without a sense of harsh self-judgment from our moral or prudential rational faculties. But, says Kant, this seeming "antinomy" of reason is not absolute. Although we can never rationally justify a selfish defection from morality, we can choose morality if we believe that, despite worldly appearances, obedience to moral duty and happiness are always aligned. This may be true if a morally wise supreme ruler of the world exists who is able to bend nature to his purposes. In that case, my sacrifice of happiness now in the name of moral integrity may be more than compensated in some future state, in or beyond the world. A religious belief of this sort is theoretically permissible, Kant argues, since, as his

[11] For a more complete discussion of these issues, see my *Religious Reason: The Rational and Moral Basis of Religious Belief* and my *Religion and Moral Reason: A New Method for Comparative Study*.

epistemological work in the first *Critique* had shown, we have no knowledge, either for or against, such a supersensible agency.

Since moral obedience itself is not strictly required by reason, it is obvious that such morally inspired religious belief is also not rationally required. Kant describes it as "a voluntary decision of our judgment ... itself not commanded" by reason (KPV 151 [146]). Nor does this belief amount to knowledge, since it reaches beyond the realm of supersensible experience and assumes a supreme moral causality whose existence cannot be demonstrated. In Kant's words, this belief constitutes a moral "faith" that can "often waver even in the well disposed but can never fall into unbelief" (KPV 151 [146]). Kant concludes the second *Critique* by suggesting that it is even a good thing that this morally stimulated religious belief does not rise to the level of knowledge. If we were assured of God's reality and power, he says, we would lose the opportunity for genuine moral commitment:

> [I]nstead of the conflict which now the moral disposition has to wage with the inclinations and in which, after some defeats, moral strength of mind may be gradually won, God and eternity in their awful majesty would stand unceasingly before our eyes.... Thus most actions conforming to the law would be done from fear, few would be done from hope, none from duty. (KPV 152 [147]; cf. LPT 123)

In the Preface to the second edition of the *Critique of Pure Reason*, Kant states that his purpose in that work is "to deny *knowledge*, in order to make room for *faith*" (KRV 29). He accomplished this by demolishing the traditional rational proofs of God's existence in that first *Critique*, and by developing the moral bases of rational faith in the second. If space allowed, I would try to indicate how influential in Kierkegaard's thinking Kant's project was. By moving religious belief out of the realm of epistemology, where theology had confined it for centuries, and by grounding it in the realm of morality, Kant helped make possible Kierkegaard's efforts to move faith away from philosophical "speculation" and re-center it in the ethical-religious existence sphere.

But my purpose here is not to trace all the lines of influence between Kant and Kierkegaard. What is most important about Kant's argument in the second *Critique* is that it leads directly to his conception of radical evil in the *Religion*. Having exposed how free we are in each moment of choice to abandon the moral point of view, Kant now realizes how fatal the implications of this insight are for our moral self-regard. Coaxed away from adherence to the moral law by the innocent but constant tug of our human inclinations to happiness, free even from rational necessitation to be moral, and aware that even a single departure from strict obedience to that law is morally unacceptable, each one of us is forced to conclude that we are radically evil: evil at the root of our willing (Rel 30).

A key component of Kant's argument here in the *Religion* is the insight that moral self-judgment will not permit a single, free departure from strict obedience to the moral law. In order to judge myself as morally good, I must *always* obey the moral law without exception. Even a single departure from moral obedience must lead me to the conclusion that I am totally lacking in moral worth. In the opening pages of the *Religion*, as he develops this point, Kant questions whether this disjunction is correct: whether we might not assert that man is by nature neither good nor evil but is at once both, in some respects good, in other respects evil. He adds, "Experience actually seems to substantiate the middle ground between the two extremes" (Rel 18).

But Kant rejects this mode of assessment. "It is ... of great consequence to ethics in general to avoid admitting, so long as it is possible, of anything morally intermediate, whether in actions ... or in human characters" (Rel 18). Behind his adoption of what Kant himself acknowledges is a "rigorist" view, lies a more basic moral apprehension. Since for Kant moral choice is always free and can never be thought of as necessitated by any inner or outer constraints, someone who has even once chosen to defect from the moral law reveals an underlying ground of choice (what Kant calls a maxim) that is not firmly committed to morality. The underlying maxim may be to respect morality *in most cases* while reserving the right to defect when moral obedience becomes too "costly" in terms of personal happiness. However, just as the promise of someone who crosses his fingers behind his back is really no promise at all, so a contingent commitment to morality is no commitment. Honest self-estimate thus imposes the conclusion that if I have even once defected from the moral law (and who, over the course of a life can assert that he has not?), I must regard my underlying moral disposition as worthless.

Kant draws from this analysis the most unsparing conclusions for human being's estimate of themselves and for their self-conceived place in a morally governed universe:

> Now this moral evil (transgression of the moral law, called SIN when the law is regarded *as a divine command*) brings with it endless violations of the law and so *infinite* guilt.... It would seem to follow, then, that because of this infinite guilt all mankind must look forward to *endless punishment* and exclusion from the kingdom of God. (Rel 66)

In the *Religion* and *The Conflict of the Faculties* this harsh conclusion leads Kant to a renewed and prolonged examination of the concept of divine grace and its role in human moral redemption. Since human beings seem to be hopelessly mired in sin, with each new choice only an occasion for further moral wrongdoing, where can they find the resources to enable them rationally to renew moral efforts? Kant answers that an admission of the possibility of divine assistance in the completion of our moral efforts is rationally allowable and perhaps rationally necessary. But even as he concedes this, Kant insists that we should put aside speculation on the how and why of grace and concentrate instead on constant rededication to moral obedience. The very stirrings of the moral law within us, and our internal call to renewed effort, he concludes, are all the signs or knowledge of grace we need. In a reflection of his deep rationalist opposition to religious dogmas and "ecclesiastical faith," Kant insists that "Action must be represented as issuing from man's use of his own moral powers, not as an effect [resulting] from the influence of an external, higher cause by whose activity man is passively healed" (CF 73). He particularly rejects emphasis here on the role of Jesus Christ as an historical figure in our redemption. Although Christ may represent the "archetype" of a human nature that is well pleasing to God, and hence a rationally conceivable possibility for us (Rel 54), the idea of the Godhead actually becoming one with a single human being is not only unnecessary but morally undesirable: "For if we think of this God-man ... as the Divinity 'dwelling incarnate' in a real man and working as a second nature in him, then we can draw nothing practical from this mystery: since we cannot require ourselves to rival a God, we cannot take him as an example" (CF 67).

Kant's analysis of the relationship between ethics, human sin, and redemption in the *Religion* and *Conflict of the Faculties* has a pronounced influence on Kierkegaard. In terms of appropriation, this influence finds expression in Kierkegaard's repeated affirmations of the "ideality" of ethics and the connection between ethics and the awareness of sin. In the *Concept of Anxiety*, for example, we read,

> Ethics points to ideality as a task and assumes that every man possesses the requisite conditions. Thus ethics develops a contradiction, inasmuch as it makes clear both the difficulty and the impossibility.... The more ideal ethics is, the better. It must not permit itself to be distracted by the babble that it is useless to require the impossible. For even to listen to such talk is unethical.... (CA 16f.)

These ideas, of course, have a long pedigree, going back to St. Paul's treatments of the law, but if we keep in mind that Kant rigorously developed a powerful theory of ethics in his concept of the categorical imperative and had shown how this concept leads to human moral self-condemnation, we can see its importance in Kierkegaard's intellectual formation. Here was the leading moral philosopher of his age (and perhaps of all times) providing rigorous logical proofs of a key doctrine of Christian orthodoxy.

If Kierkegaard clearly profited from Kant's analyses of ethics and sin (and there are many other signs in his writings that he did so[12]), he could not agree with Kant's solution to the problem: self-initiated moral renewal that is conceptually indifferent to its grounding and possibility and that ignores the role of God's active redemptive activity on our behalf. Not only did Kierkegaard actively resist Kant's rationalistic presentation of the incarnation in the *Fragments* and elsewhere, but also, in sharp disagreement with Kant, he stresses the absolute importance of God's intervention into history to renew our wills by providing us the "condition" for our moral renewal.

Remarkably, Kierkegaard did not merely disagree with Kant. He appears, in two important ways, to have drawn on Kant's own arguments to defeat him. We can only surmise and reconstruct the first of these efforts: Kierkegaard's possible perception of the poverty of Kant's arguments as a basis for constructing his own defense of the idea of the importance of an historical savior. I have already touched on aspects of the Kantian arguments, but let me sketch them more fully here. Kant believed that while divine grace

[12] There is substantial consensus among Kierkegaard scholars that Kantian thinking deeply influences the ethical position developed by Kierkegaard in *Works of Love* and *Purity of Heart*. For discussions of this, see Jeremy D.B. Walker, *To Will One Thing: Reflections on Kierkegaard's 'Purity of Heart'* (Montreal, Canada: McGill-Queen's University Press, 1972) and M. Jamie Ferreira, *Love's Grateful Striving: A Commentary on Kierkegaard's Works of Love* (New York: Oxford University Press, 2001).

may be necessary to accomplish our moral rebirth, we are better off not dwelling on how this is accomplished. Such speculation not only takes us beyond the realm of possible human knowledge but it also raises profound and perplexing moral questions. For example, how can we bring our conception of moral freedom and accountability together with the idea of a divine agency that enters into and transforms our will? This idea is not impossible, since moral freedom and causal determination (which we must ascribe to ourselves as phenomenal creatures) co-reside in every instance of human willing. But it is best not to engage in such transcendental inquiries if we do not have to. All we need for moral renewal is our conscience and inner commitment to reform, which we may take as signs of a divine presence and whose operation we need not further understand.

Kant adds to this conception of grace a deliberate rejection of any role for a historical savior in its accomplishment. We saw that he believed that such a figure could only serve as a negative example for us since the union of God and man would create a person beyond the sphere of our own moral experience and not subject to our frailties. Should God dwell in a human being (something Kant's epistemology does not strictly rule out since the union of the eternal and temporal is an aspect of our own moral willing), we can also never *know* that this is the case because our knowledge is always confined to phenomena and appearances. Finally, in a remark I have already partially quoted in connection with Kant's reply to objections against his rationalist conception of grace, Kant states that God's entry into time to offer us salvation is unnecessary, because

> ... there is no other way we can conceive the decrees of a holy and benevolent law-giver with regard to frail creatures who are yet striving with all their might to fulfill whatever they recognize as their duty; and if, without the aid of a definite, empirically given promise we have a rational faith and trust in His help, we show

> better evidence of a pure moral attitude and so of our receptivity to the manifestation of grace we hope for than we could by empirical belief. (CF 83)

The weakness and inappropriateness of Kant's arguments here are glaring, and surely must have seemed so to Kierkegaard. Having developed the depths of human sinfulness in the *Religion* and the awareness that even one act of wrongdoing can point to a corrupt underlying will, Kant now offers more of the same moral striving as the remedy for corruption and as the sign of divine grace. But if every single act of willing is tainted, why should we think better of these renewed (and possibly self-deceiving) stirrings? We are reminded here of Kant's own assertion in the *Religion* that "Man is never more easily deceived than in what promotes his good opinion of himself" (Rel 62). Worse, there is a degree of presumption in Kant's position that is morally offensive and self-condemning. It amounts to saying that because I need grace to overcome my own iniquity, I can be sure that I will receive it. The imaginary rationalist philosopher whose objection Kant had sketched and to whom the remark above was meant as a reply makes this clear in his objection when he says:

> To believe that God, by an act of kindness, will in some unknown way fill what is lacking to our justification is to assume gratuitously a cause that will satisfy the need we feel (it is to commit a *petitio principii*); for when we expect something by the grace of a superior, we cannot assume that we must get it as a matter of course; we can expect it only if it was actually promised to us.... (CF 83)

Far from answering this objection, Kant has only shown his insensitivity to its force.

Finally, Kant's assertion that "we are better off morally by not having confidence in the definite, empirically given promise" of grace because it heightens our rational faith and trust in God and evidences "a pure moral attitude" (CF 83) represents an inappropriate application of one of Kant's better insights. We encountered this insight at the end of the second *Critique,* where it was used to illuminate the value of our cognitive distance from God's agency in rewarding virtue and, hence, our need to choose morally without any certainty that our actions will lead to our happiness. In that instance, religious uncertainty purifies moral striving. But in the case at issue here, where we confront our bondage to sin and justified eternal perdition, of what benefit is the absence of any concrete token of God's mercy? How does God's absence here purify or renew moral striving? If anything, it would seem that this is one place where our whole moral consciousness cries out for some concrete sign of divine support.

A careful reading of the *Religion* and *The Conflict of the Faculties,* then, would certainly have inspired Kierkegaard to an ardent defense of traditional Christian concepts. Indeed, in Kant's arguments in these two books, which Kierkegaard presumably pored over in preparation for his *Attestats* examination, he would have found many of the conceptual timbers he needed to begin construction of his own religious authorship.

A second (and epistemological) aspect of Kant's arguments might also have served Kierkegaard well. Kant's rhetorical debate between himself and a rationalist objector in the *Conflict of the Faculties* reveals a deeper problem in Kant's position on grace. The objector, we saw, indicates that there is an important difference between expecting that something be given to us and actually getting it. Otherwise said, it is one thing rationally to postulate an occurrence as conceptually necessary for us, but it is quite another to say that that occurrence has happened and that the state of affairs to which it points actually *exists.*

The difference between logical necessity and real existence (the givenness of something in our experience) is, of course, at the heart of Kant's rejection of the ontological and teleological arguments for God's existence in the first *Critique.* There, Kant had shown that existence cannot be treated as a predicate in such a way that something is made more perfect by possessing it: by existing rather than not existing. The predicative use of existence was at the heart of the classical ontological argument, where God, the all-perfect being who possesses all predicates of value, was said necessarily to have the attribute of existence. But the use of existence as a predicate is wrong, says Kant. Existence is "a copula of a judgment" that affirms the object's presence, with all its predicates, in our experience. In any description of a thing, Kant tells us:

> The small word "is" adds no new predicate, but only serves to posit the predicate *in its relation* to the subject. If, now, we take the subject (God) with all its predicates ... and say "God is," or "There is a God", we attach no new predicate to the concept of God, but only posit the subject itself with all its predicates, and indeed posit it as being an *object* that stands in relation to my *concept*.... Otherwise stated, the real contains no more than the merely possible. A hundred real thalers do not contain the least coin more than a hundred possible thalers.... My financial position is, however, affected very differently by a hundred real thalers than it is by the mere concept of them.... (KRV 505)

A bit further on, Kant adds,

> If we think in a thing every feature of reality except one, the missing reality is not added by my saying the defective thing exists. On the contrary, it exists with the same defect with which I have thought it, since otherwise what exists would be something different from what I thought. (KRV 505)

Readers of Kierkegaard should have no problem identifying his debt here to Kant. Arguments about the difference between concept existence and real existence abound throughout Kierkegaard's writings. In the "Interlude" of the *Fragments,* we find whole paragraphs that evoke Kant's arguments. Thus, in a passage resonant with the Kantian language, Climacus tells us that "[I]f that which comes into existence does not itself remain unchanged ... then the coming into existence is not *this* coming into existence" (PF 73). Even Kant's summary quip about the hundred dollars (a quip that Kierkegaard clearly appreciated because he cites it several times in his writings (PAP VIII-2 B 81 [JP1–649]; PAP VIII-2 B 82 [JP1 650]; PAP X1 A 666 [JP3–3558]) is alluded to when Climacus resumes his own position by saying, "A fly, when it is, has just as much being as the god" (PF 41n.).

I will not rehearse arguments that I have made elsewhere that the "Interlude" of the *Fragments* is absolutely permeated by Kantian epistemological points.[13] The more interesting question is why Kierkegaard feels compelled at this point in his authorship to return so explicitly to Kant in the course of an argument about history and salvation. My answer to that question is that here, Climacus-Kierkegaard is turning Kant's epistemology against Kant's denial of the need for a historical savior in the economy of our salvation. The first part of the *Fragments* develops the "Socratic" position that we have everything we need to accomplish our own redemption: the teacher is only the "occasion" for our moral learning. This is precisely the view Kant defends in the *Religion* and the *Conflict*. Having knowledge of our sin, we also *know* that we may reasonably presume the divine assistance required for overcoming it. In other words, when we properly understand the concepts of ethics, we at once understand sin and the redemption from sin, because as "frail creatures striving with all their might to fulfill what they recognize as our duty," there is "no other way we can conceive the unyielding decrees of a holy and benevolent law-giver" (CF 83) than to believe we might be eligible to receive God's grace.

---

[13] Ronald M. Green, "Kierkegaard's Philosophical Fragments: A Kantian Commentary" in *International Kierkegaard Commentary*, in *Philosophical Fragments and Johannes Climacus*, ed. Robert L. Perkins (Macon, Georgia: Mercer University Press, 1994) 169–202.

But Kant is wrong. As his rationalist objector observes, there is an enormous difference between expecting something by the grace of a superior, and actually getting it. Between the two lies the difference of existence. Have we or have we not been redeemed? Has God's ability to act on his presumably benevolent will really been manifest in our experience? Has it been manifest in history? Remarkably, Kant, for all his insistence on the sheer givenness and non-necessity of existence seems to miss this point, although Kierkegaard has not. The long epistemological "Interlude" of the *Fragments* calls Kant to account on the basis of Kant's own philosophy.

A small, but especially interesting sign of this is Climacus-Kierkegaard's use of the term *accessorium* in a preceding discussion of the concept of "existence" in the *Fragments* (PF 40). Although there is no reference to Kant in the text, Kierkegaard's papers show that he knowingly borrowed the term from Kant.[14] But where in Kant? Remarkably, Kant apparently never uses this term in his epistemological writings. But he does use it, as we saw above, in *The Conflict* to identify the historically given, and hence not rationally necessary elements of Scripture (CF 117). Hence, the single use of this term provides evidence that Kierkegaard associated Kant's epistemological arguments about existence with Kant's treatments of the question of historical revelation.

The issue goes deeper still. A close reading of Kant's *Religion*, and his analysis of radical evil, indicates the importance of phenomenal experience—experience in time—for Kant's understanding of sin. Kant is very clear that our root propensity to evil

[14] "I never reason in conclusion to existence (for in that case I would be mad to want to reason in conclusion to what I know), but I reason in conclusion from existence and am so accommodating to popular opinion as to call it a demonstrative argument. Thus the connection is somewhat different from what Kant meant—that existence is an *accessorium* [addition]—although therein he undeniably has an advantage over Hegel in that he does not confuse."—PAP V B 5:3 (PF, Supplement, 190).

does not derive from the fact that we are phenomenal creatures with finite needs and inclinations. These inclinations (the sum of whose satisfaction constitutes our happiness) can as easily lead to moral obedience as disobedience. No, Kant says, there is nothing necessary about sin. It derives from the exercise of our freedom: from a choice identifiable from the earliest exercises of our wills to place self over others, the particular over the universal.

Kant's rejection of the necessity of sin means that in each moment of choice we remain free to choose for or against morality. Experience shows that we sometimes choose immorally, and, as we have seen, even a single instance of mischoice, because of its possibly eternal import as evidence of the underlying maxim of all our choices, is fatal for our self-estimate. But even though a single instance of choice potentially has timeless consequences, it still occurs in time. It is choice in time, whether for good or for evil, that dictates our moral destiny.

It follows from this that choice in time must be equally important for our moral salvation. For if that salvation could be developed out of necessary concepts, then the original choice that determined it would lose significance. We might be constituted so as to be capable of freely defecting from morality, but that defection would be inconsequential because it would be eternally remedied by our Creator's necessary provision of moral redemption. That this is not the case, that the Creator has graciously chosen to enter time and history to redeem us, tells us that he could have done otherwise. As a result, our free choice remains infinitely consequential because it could lead to redemption or perdition. In contrast, the rationally necessary redemption that Kant argues for radically minimizes the importance of our wrongful choice and suggest that these choices make really no difference in the shape of our final moral destiny. Thus, our moral freedom requires God's freedom, and God's freedom is shown in the givenness and non-necessity of his redemption of us in history. Whether God has in fact entered history in this way, therefore, is of absolute importance for our eternal salvation. But, of course, these ideas are among the leitmotifs of Kierkegaard's whole authorship.

There is so much more that could be said here. If the first part of the *Fragments* deals with the Kant's Socratic challenge, and the "Interlude" invokes his epistemology and ethics against him, the third part, dealing with the contemporary believer relies heavily on Kant's understanding that sense experience can never itself authenticate ethical and religious truths. The twist, of course, is that the truth that Kierkegaard wishes to defend is not an eternal logical necessity but the reality of the god in time and the unique bestowal by the god of "the condition" that redeems. Once that happens and once that historical fact is accepted, however, Kierkegaard himself tells us, "everything is again structured Socratically" (PF 65). Empirical/historical information about the redemptive event and even historical proximity to it do us no good, since, in Kantian terms, it is the conceptual act of faith, not sensible experience that counts. As Kant tells us "if God should really speak to man, man could still never *know* that it was God speaking. It is quite impossible for man to apprehend the infinite by his senses, distinguish it from sensible beings, and *recognize* it as such" (CF 115). Thus, the *Fragments*, at least, comes full circle: beginning as an argument against Kant, turning Kant's own positions against him, it closes with a return to a transformed but still deeply Kantian assumption about primacy of concepts in judging experience. The difference is that Kant's concepts are self-developed, whereas Kierkegaard's are bestowed in time by the god.

I will stop here. I am confident that a reader who closely examines all of Kierkegaard's pseudonymous writings following a careful reading of Kant's two *Critiques*, the *Religion* and *The Conflict of the Faculties* will be impressed by the range and extent of borrowings that extend from simple terms to whole arguments. This debt is evident despite the nearly total absence of explicit references to Kant. As many students of Kierkegaard have noted, it was typical of him for many reasons to ignore or obscure his sources. Sometimes, he even used stalking horses when other unnamed but more relevant sources are in mind.[15] What remains important is the fact that Kierkegaard was one of Kant's best nineteenth-century readers. In creating his authorship, Kierkegaard relied on key elements of Kant's epistemology, ethics, and philosophy of religion. These elements were picked up, transmuted and used by Kierkegaard to reinforce his own Christian convictions.

---

[15] See Jon Stewart, *Kierkegaard's Relations to Hegel Reconsidered* (New York and Cambridge: Cambridge University Press, 2003) 388.

6

"Developing" *Fear and Trembling*

"Once I am dead, *Fear and Trembling* alone will be enough for an imperishable name as an author. Then it will be read, translated into foreign languages as well (PAP X2 A 15 n.d., 1849).

Kierkegaard was prophetic in his estimate of the place of *Fear and Trembling* in his authorship. Although several of his pseudonymous works have also become philosophical classics, *Fear and Trembling* continues to haunt us like no other writing of his. Its defense of individual existence still resonates at the end of a century marked by horrifying mass movements, while its depiction of radical religious obedience stirs new fears as we enter a period when older political ideologies are being replaced by renewed expressions of religious absolutism.

Fear and Trembling remains so evocative partly because of its enigmatic nature. From the outset—by means of the famous epigraph drawn from Johann Georg Hamann—Kierkegaard signals that not everything that follows is as it seems. Beyond this, there is evidence that Kierkegaard designed *Fear and Trembling* as a text with hidden layers of meaning. In *The Point of View for my Work as an Author,* Kierkegaard tells us that the most important ethical and religious truths cannot be communicated directly, as though one were writing on a blank sheet of paper. Instead, they demand creative endeavor by the author and a corresponding effort by the reader that involves "bringing out by means of chemicals some writing that is hidden under another writing" (POV 54). Kierkegaard appears to have in mind the process by which a message written in secret ink is deciphered.[1] Although he wrote before the advent of modern photography, we can also think of his interpretive advice in terms of the process whereby film is developed. Beginning with a surface material of a certain texture and color, we undertake to expose different latent images and ideas. In what follows, I want to take Kierkegaard's advice and "develop" the text of *Fear and Trembling* in this photographic sense. What we will see, I think, is that this text contains not just two, but multiple levels of meaning. Each level has its own significance, and as we expose each deeper level, the messages grow more subtle. Finally, when development is complete, we have in our hands a transparent image through which we can see all the levels of meaning and which, when held to the light, reveals a religious-ethical communication of surprising richness and complexity.

Level One: The Call to Christian Commitment

At the first and most apparent level of meaning, *Fear and Trembling* is a stinging critique of both the popular and cultured Christianity of his day and a reminder of the primitive challenge of Christian faith. This critique is signaled by the choice of Abraham and the Genesis 22 episode as paradigms of faith and by the repeated use of commercial metaphors to portray the spirit of the age.

Kierkegaard believed that the cultural triumph of Christian civilization had effaced the primitive meaning of Christianity. A religious identity, whose acquisition once entailed great risk, had become a matter of merely being born to Christian parents in a

---

[1] For an analysis of *Fear and Trembling* in terms of this image, see my "Deciphering *Fear and Trembling*'s Secret Message," *Religious Studies* 22 (1986): 95-111. The choice here of the metaphor of photographic development signals my more recent view that *Fear and Trembling* contains not one but a series of interrelated manifest and latent meanings.

Christian nation. Time had also transmuted the stories of the early heroes and saints of faith. Looked at with the benefit of eighteen centuries of hindsight, a false picture of historical success and well-earned veneration had replaced vivid experiences of individual risk, suffering, abandonment, or martyrdom. What Kierkegaard calls "the results" had come to overshadow the anguished choices by early Christian disciples.

As an antidote to this spiritual lethargy, Kierkegaard's pseudonym, Johannes de Silentio, devises what amounts to a theological shock treatment.[2] He portrays Abraham, the "knight of faith" in the full terror of his encounter with the divine command. By following the patriarch step-for-step on his difficult journey to Mount Moriah, Johannes seeks to recover aspects of faith that years of saintly veneration and familiarity with the happy ending had effaced. "What is omitted from Abraham's story is the anxiety," Johannes tells us (FT 28). "We are curious about the results, just as we are curious about the way a book turns out. We do not want to know anything about the anxiety, the distress, the paradox" (FT 63).

To illustrate how far the Christianity of his day had erred from the primitive experience of faith, the "The Preliminary Expectoration," portrays an imaginary churchgoer who is led by a preacher's sermon to want to imitate Abraham. Learning of this, the pastor visits the parishioner and, rising to unprecedented heights of rhetorical fervor, thunders, "You despicable man, what devil has so possessed you that you want to murder your own son?" (FT 28). In Johannes's view, it does this hapless fellow little good to reply, "But, after all, that was what you yourself preached about on Sunday," since the established church and its functionaries were ignorant of how thoroughly they had replaced primitive Christianity with a cliché-ridden, worldly piety.

---

[2] Paul Dietrichson, "Kierkegaard's Concept of the Self," *Inquiry* 8, no. 1 (Spring 1965): 2.

The use of Abraham also conveys a new emphasis on faith as a way of life. This emphasis is meant to replace the centuries-old understanding of faith as merely an acceptance of dogmatic truths. Abraham is a fitting choice to communicate this lesson because his hallmark is not intellectual achievement, but a prodigious ability to live trustingly and obediently. In the margin of a draft of the "Eulogy on Abraham," Kierkegaard makes this point even clearer by ending the section with a definition of faith "not as the content of a concept but as a form of the will" (PAP IV B 87:2 n.d., 1843). The emphasis on willing and acting rather than thinking or reasoning is also highlighted by the sheer irrationality of Abraham's faith, his belief "by virtue of the absurd" that he will get Isaac back. As Jerry Gill points out, to present a "dialectical corrective," Kierkegaard offers the story of Abraham as *a reductio ad absurdum* of all traditions that see faith as involving mental assent.[3]

These ways of evading religious-ethical commitment represent pervasive and abiding problems in Christianity, but, in Kierkegaard's day, evasion had taken a new and virulent form. Among the intellectual leadership and scholarly teachers of Christianity, a pseudo-Christianity permeated by Hegelian philosophy held full sway. Under the motto "One must go further," the Hegelians presented "faith" as a rudimentary phase of intellectual development to be transcended by their own rational philosophy. This philosophy radically subordinated matters of personal ethical and religious decision, the crucial events of individual history, to scholars' comprehension of the meaning of world history. For the Hegelians, Abraham was, at best, a figure of historical interest whose personal trial and response was unimportant compared to whatever historical

---

[3] Jerry H. Gill, "Faith Is as Faith Does," in *Kierkegaard's* Fear and Trembling: *Critical Appraisals,* ed. Robert L. Perkins (Tuscaloosa: University of Alabama Press, 1981) 204.

significance he might have in the development of monotheism. It is to question this approach that Johannes takes us step by step on Abraham's arduous journey.

Jibes at the Hegelians also virtually bracket the text. The book begins, for example, with Johannes remarking that "Not only in the world of business but also in the world of ideas, our age stages *ein wirklicher Ausverkauf* [a real sale]," in which everything can be had at "a bargain price" (FT 5). Once, faith was a task for a whole lifetime, but now, "every speculative monitor who conscientiously signals the important trends in modern philosophy, every assistant professor, tutor, and student," is unwilling to stop even with doubting but "goes further."

Near the book's end, the epilogue returns to business matters. Johannes mentions the practice of merchants in Holland sinking cargoes of spices in the sea to jack up declining prices. This use of the language of commerce highlights the era's (and the Hegelians') bourgeois preoccupations and mocks the age's enthusiasm for mass-produced, bargain-priced faith. Translated to the realm of spirit, the Dutch merchants' practice proves instructive to Johannes. Just as they sacrificed their cargoes to raise the value of their goods, so he employs the dramatic story of Abraham's sacrifice to raise the price—and cost—of faith. Understood at this level, it is precisely the outrageousness of Abraham's conduct that makes it the fitting counterweight to cultural Christianity and Hegelian philosophy.

Level Two: The Psychology of Faith

Once we have assimilated *Fear and Trembling*'s deliberately shocking indictment of cultural Christianity, our encounter with the text exposes a less dramatic and more subtle level of meaning. At this level, *Fear and Trembling* involves an exploration of the psychology of faith. This inquiry starts with the first level's assumption that faith is a lived commitment but seeks to understand its precise mental content for the believer. In the "Exordium," "Eulogy on Abraham," and "Preliminary Expectoration," Johannes largely sets polemic aside to focus on the psychology of various exemplars of faith, some of who prove to be quite ordinary persons. Here, Johannes lets us know that what is important in faith is not outer deeds like Abraham's dramatic obedience, but quiet and difficult inner movements of the spirit.

The central idea here is the "double movement" of faith. The first movement, "infinite resignation," is accomplished by relinquishing one's heart's desire. For the young swain depicted by Johannes, who accepts the fact that the great love of his life lies forever beyond his reach, infinite resignation leads to the discovery of his "eternal consciousness." Like the shirt whose thread is spun in tears, infinite resignation provides "peace and rest and comfort in the pain" (FT 45).

The "knight of faith" embodies the second movement. He starts where the "knight of infinite resignation" ends:

> He does exactly the same as the other knight did: he infinitely renounces the love that is the substance of his life, he is reconciled in pain. But then the marvel happens; he makes one more movement even more wonderful than all the others, for he says: Nevertheless, I have faith that I will get her—that is, by virtue of the absurd, by virtue of the fact that for God all things are possible. (FT 46)

Abraham makes these two movements. He obeys God's command and willingly relinquishes Isaac. Simultaneously, he continues to believe that God will not demand Isaac of him, and that he will again enjoy his son's presence in this life. Unlike the despairing versions of Abraham presented in the *Exordium* who manage to fulfill God's will but lose the resilience of their souls, the real Abraham retains the ability "once again to be happy in Isaac" (FT 35).

Johannes lets us know that the capacity for such knighthood is not confined to the older heroes and saints of faith but remains available to every human being. He imagines a knight of faith residing in the Copenhagen of his day. No outward signs reveal this person's spiritual depth. In every way he resembles a bourgeois philistine, even a tax collector. Inwardly, however, at every moment he is making "the movement of infinity." He feels the pain of renouncing everything, "yet the finite tastes just as good to him as one who never knew anything higher" (FT 40).

Edward Mooney argues that at this level of psychological development, *Fear and Trembling* aims at describing and commending a stance of selfless care. According to Mooney,

> The knight of faith can both renounce and enjoy the finite because he sees, or knows in his bones, that renouncing all *claim* to the finite is not renouncing all *care* for it. He is at home and takes delight in the finite (witness the tax-collector) because he cares; yet this is a selfless care, for he has given up all proprietary claim.[4]

If Mooney is right, this level of meaning of *Fear and Trembling* begins to suggest to us that the text as a whole is not quite the terrifying defense of religiously commanded homicide it seems to be. Rather, it begins to appear as a more traditional defense of selfless love as a central feature of the religious life.

Level Three: The Normative Shape of Christian Existence

If a first level of meaning contains a call to strenuous lived commitment to Christian faith, and a second develops the psychology of faith and love, a third level of *Fear and Trembling* explores the question of the norms that should guide the conduct

---

[4] Edward Mooney, "Understanding Abraham: Care, Faith, and the Absurd," in Kierkegaard's *Fear and Trembling: Critical Appraisals*, ed. Robert L. Perkins (Tuscaloosa: University of Alabama Press, 1981) 108.

of a committed Christian. At this level, *Fear and Trembling* appears to be at least the beginnings of a study in ethics.

This normative inquiry comes to fore in the three "Problemas," especially in the ideas of a "teleological suspension of the ethical" and an "absolute duty to God." These difficult conceptions have been interpreted in various ways, not all of which are consistent with one another, nor with some of Johannes de Silentio's own claims and statements. Part of the problem may be due to faulty readings of the text. Others seem to reflect Johannes's complex and confusing position. As we will see, these difficulties ultimately drive us beyond ethics to a still deeper level of meaning of the book.

The first "Problema" presents Genesis 22 as involving "a teleological suspension of the ethical" in which Abraham, the knight of faith, subordinates his responsibilities as a father to the needs of his own personal relationship with God. The ethical, Johannes tells us, is the "universal" and it is one's ethical responsibility to annul one's singularity to "become the universal" (FT 54). Although the Kantian and Hegelian philosophies Johannes presumes develop this thought in complex ways, the idea expressed here amounts to the simple requirement that the needs of the common good take precedence over merely individual wishes. But as it is exemplified in Abraham, faith reverses this priority. Faith is "the paradox that the single individual is higher than the universal" (FT 55).

Johannes insists that we cannot comprehend Abraham's behavior in ethical terms. From the point of view of ethics, Abraham is nothing more than the murderer of his son. At no time is he a "tragic hero" who sets aside one expression of the ethical for a still higher expression. Agamemnon, Jeptha, and Brutus are tragic heroes. They also willingly undertake to kill their children, but as responsible leaders, they do so to protect the welfare of their community and the common good. Not Abraham. "By his act he transgressed the ethical altogether and had a higher *télos* outside it in relation to which he suspended it.... Why, then, does Abraham do it? For God's sake and—the two are wholly identical—for his own sake" (FT 59).

Johannes's description of Abraham poses a sharp challenge to those who would make sense of *Fear and Trembling* as a study in ethics. On the one hand, Johannes does not shrink from depicting Abraham as fully outside the ethical—as truly the murderer of his son. Not only does his conduct violate one of our most important ethical norms, it cannot be rationally justified in any way. His conduct remains "for all eternity a paradox, impervious to thought" (FT 56). On the other hand, Johannes also frequently lauds the patriarch and holds him up as a model for the Christian life. "I cannot understand Abraham," he tells us, "I can only admire him" (FT 112; Cf. 57, 114). Reading *Fear and Trembling* as a work intending to offer at least a preliminary vision of the Christian moral life produces a jarring inconsistency. *Fear and Trembling* seems to hold up as exemplary and somehow worthy of imitation a kind of conduct that we cannot possibly encourage, defend, or understand in terms of general moral values.

Various interpreters have tried to reduce or eliminate this seeming contradiction. Elmer Duncan, for example, argues that the primary target of Johannes's ethical critique is a kind of Kantian ethical absolutism that makes no room for permissible exception cases. Kant, in his essay, "On a Supposed Right to Lie from Altruistic Motives," had argued that one is not permitted to tell even a small lie to a criminal aggressor in order to save the life of an innocent person.[5] Duncan believes that Kierkegaard, like many other readers of Kant, found this position to be preposterous. Since he was unable to justify exceptions within the rigid theoretical framework established by Kant, he was compelled to locate their possibility in the religious domain and to

[5] KPR 346–350.

argue for the superiority of the religious over the ethical sphere of life. Duncan concludes by dismissing what he takes to be the argument of *Fear and Trembling* by pointing to other, less radical ethical approaches to the problem of exceptions.[6]

This line of interpretation is interesting but it poses at least two problems. First, it is not clear that it is Kantian absolutism that Kierkegaard has in mind in framing the "teleological suspension of the ethical." The tragic hero as portrayed by Johannes is no absolutist. To fulfill a "higher" obligation to the state, Agamemnon, Jeptha and Brutus are willing to break the moral rule against murder. Yet, according to Johannes, each is a tragic hero, not a knight of faith, and their behavior does not involve a teleological suspension of the ethical. Second, this interpretation ignores Johannes's repeated affirmations that in suspending the ethical, Abraham moved entirely outside its sphere. There is no "higher expression for the ethical that can ethically explain his behavior" (FT 57). In view of this, it becomes difficult to construe Abraham as seeking to break away from rigid ethical confines to express a more nuanced understanding of moral obligation.

Similar problems trouble a second interpretation of the ethical position sketched out in *Fear and Trembling*. This views the book as a critique of ethical philistinism. As Gene Outka notes, those who hold this view understand Abraham as acting contrary to established public opinion. "He violates the canons of respectability and offends those who take as authoritative and moral opinions of their class and circumstance. The levels of dread and conflict he knows are out of reach of prosaic temperaments

[6] A similar reading is offered by Geoffrey Clive, who argues that the ethical that is transcended or suspended here is the morality of general rules—"The Teleological Suspension of the Ethical in Nineteenth-Century Literature," *The Journal of Religion* 34 (April, 1954): 75–87.

who are content to abide by conventional rules of their historical epoch."[7]

Outka points out that there is clearly something to this interpretation. In an impressionistic sense, it fits the general timbre of the book. It also connects well with *Fear and Trembling*'s undeniable emphasis on active, lived commitment to one's ethical or religious values. Nevertheless, by presenting Abraham as a sincere defender of genuine ethical values as opposed to mere conventionalism, this interpretation runs up against Johannes's repeated statements that Abraham's conduct is totally beyond ethical justification. This interpretation also does not fit with Johannes's important distinction between Abraham and the tragic hero. The tragic hero, Outka observes, "also requires courage and may violate conventional moral opinions."[8] Yet the tragic hero's behavior does not exemplify faith nor involve a teleological suspension of the ethical.

A third interpretation sees Johannes's argument as sharpening the book's broad critique of Hegelian philosophy. Now the focus is on Hegel's ethics, especially the primacy he places on the public morality and social roles embodied in his idea of *Sittlichkeit*. For Hegel, as for Kant, ethics involves subordinating individual inclinations to the demands of the universal. Hegel further insists that the universal in ethics must take form in the concrete public life of a people, institutionalized in family, civil society, and the state. The state itself, he proclaims, is an earthly deity [*Irdisch-Göttliches*] that commands our highest loyalties.[9]

It is easy to read *Fear and Trembling* as a critique of this Hegelian ethical position. At the opening of the first Problema,

[7] Gene Outka, "Religious and Moral Duty: Notes on *Fear and Trembling*," in *Religion and Morality*, ed. Gene Outka and John P. Reeder (Garden City, New York: Anchor Press/Doubleday, 1973) 212.

[8] Ibid., 212.

[9] G.W.F. Hegel, *Elements of the Philosophy of Right*, ed. Allen W. Wood (Cambridge: Cambridge University Press, 1991) 307 [§ 272].

Johannes refers to Hegel and asks whether "social morality" in the Hegelian sense really is the highest (FT 55). As presented by Johannes, Abraham clearly violates his two principal social role responsibilities: as a father and leader of his people. Indeed, since Isaac's life represents the promised continuance of the people, both these roles are simultaneously violated by Abraham's conduct. As Johannes tells us, "Insofar as the universal was present" for Abraham, "it was cryptically in Isaac, hidden so to speak, in Isaac's loins...." (FT 59). By subordinating these compelling communal responsibilities to his own spiritual salvation, Abraham takes a step beyond any social definition of the self.

Taken as a critique of Hegel's ethics, *Fear and Trembling* can be read in two different ways that reduce the apparent inconsistency between Johannes's moral condemnation and praise of Abraham. One the one hand, we can see *Fear and Trembling* as an ethical statement rejecting Hegel's nearly total subordination of the individual to the nation state and as a prophetic defense of the rights of the individual in the face of oppressive social collectivities. Those who read the book this way see it as an incipient protest against the horrendous totalitarian movements that nineteenth-century mass philosophies were to produce, some of which, like Marxism, were based on Hegel's thought. By affirming the priority of the individual, *Fear and Trembling* is seen as offering an important corrective to this dangerous loss of self. Unfortunately, this very common reading of *Fear and Trembling* draws its force from the implicit idea that Abraham's conduct somehow represents a higher ethical possibility than Hegel's nationalism. As such, it runs directly up against Johannes's repeated statements that Abraham cannot be ethically "mediated" or understood. The importance of the individual, and prophetic resistance to the mass, are major themes certainly present in abundance elsewhere in Kierkegaard's writings. But, unless we assume, as some interpreters have done, that Johannes has merely resorted to hyperbole in denying the moral justifiability of Abraham's conduct,[10] it violates the spirit of *Fear and Trembling* to read Johannes's defense of Abraham primarily in these terms.

Some who read the book as a rejection of Hegel's ethics take it not so much as dealing with moral rules and appropriate forms of conduct, as a call for personal individuation. Jerome Gellman puts this well when he states that the "voice of God," for Kierkegaard, is not literally a command to do a specific act. Rather, it is

> a "call" out of the "infinity" of the self, for self-definition as an individual, as opposed to self-definition from within the institutions of society, specifically the family.... The story is not about Abraham's daring to kill his son, but is about Abraham's having the courage to be willing to see himself not as a father, but as an individual.[11]

This interpretation has the advantage of squaring well with the text's obvious critique of bourgeois complacency and the Hegelians' own repeated tendencies to smooth down the hard edges of faith and ethics. It has the disadvantage of leaving the normative ethical import of *Fear and Trembling* in doubt and of leaving us wondering why, apart from its shock value, Genesis 22, of all biblical texts, was chosen to make this point.

A final major interpretation of the normative level of *Fear and Trembling* sees Johannes (and perhaps Kierkegaard) as unabashedly defending a "divine command" view of ethics. Those who read the book this way maintain that the lesson of Abraham's conduct is that every committed religious person must remain open to the possibility of a direct command from God that takes precedence over any rational ethical duties. Supporting this interpretation are Johannes's clear repudiation at the outset of the

---

[10] George D. Chryssides, "Abraham's Faith," *Sophia* 12 (April 1973): 10–16.

[11] Jerome Gellman, "Kierkegaard's *Fear and Trembling*," *Man and World* 23 (1990): 297.

second Problema of the Kantian ethical position that denies there are any direct duties to God and that sees all relationship to God as contained within obedience to the rational moral law (FT 68). Also supporting it are several explicit statements in *Fear and Trembling* that for Abraham "duty is simply the expression for God's will" (FT 60; Cf. 70).

Some who defend this view of the book's ethical message see it as solving the puzzle of how Johannes can both paint Abraham in the starkest ethical terms and laud him as a model of religious behavior. The solution lies in Kierkegaard's/Johannes's assumptions about God's nature. Johannes tells us early on, for example, that he is convinced that "God is love" (FT 34). Within the context of such a belief, unstinting obedience to God makes sense even when he appears to require horrific deeds or sacrifices, as in the case of Genesis 22.

C. Stephen Evans develops an interpretation of this sort when he argues that it is Abraham's "special relationship with God" that explains his inability to offer reasons for his conduct while still maintaining his ethical integrity and moral resolve. Because of this special relationship, says Evans,

> Abraham knows God as an individual; he knows God is good, and he loves and trusts God. Although he does not understand God's command in the sense that he understands why God has asked him to do this or what purpose it will serve, he does understand that it is indeed God who has asked him to do this. As a result of his special relationship, Abraham's trust in God is supreme. This trust expresses itself cognitively in an interpretive framework by which he concludes, all appearances to the contrary, that this act really is the right thing to do in this particular case. God would not in fact require Isaac of him...; or even if God did do this thing, he would nonetheless receive Isaac back.... Abraham's willingness to sacrifice Isaac might be compared with the

confidence of a knife-thrower's assistant in the accuracy of a knife-thrower's aim.[12]

This type of divine command position—one combining what Outka calls "general trust with specific perplexity"[13]—has a venerable place in the traditions of commentary on Genesis 22.[14] It certainly corresponds well with Kierkegaard's own personal religious position. In various writings, he unites a firm insistence on God's unwavering goodness[15] with an aversion to autonomous Kantian ethics and a preference for a divine command position. At several points, Kierkegaard tells us that to view moral requirements as self-imposed can only lead to moral laxity. For our moral obligations to elicit our full respect, we must regard them as emanating from an authoritative lawgiver who issues and upholds his commands (PAP X2 A 396 n.d., 1850 [JP1–188].

It may be that this is the ethical position Kierkegaard wished to present in *Fear and Trembling* and for which he has Johannes employ the example of Abraham's conduct in Genesis 22. But if so, *Fear and Trembling* is strangely lacking in the development of such a view. Johannes places great emphasis throughout on the horrifying nature of Abraham's conduct and the willingness of a "knight of faith" to go beyond ethics. But there is very little mention of the theological beliefs needed to render this view ethically comprehensible or compelling. Apart from the one fleeting remark about God's love in the "Preliminary

[12] C. Stephen Evans, "Is the Concept of an Absolute Duty toward God Morally Unintelligible?" in *Kierkegaard's* Fear and Trembling, ed. Robert L. Perkins (Tuscaloosa: University of Alabama Press, 1981) 145.

[13] Gene Outka, "Religious and Moral Duty: Notes on *Fear and Trembling*," in *Religion and Morality*, 243.

[14] See my *Religion and Moral Reason*, chapters four and five, and David A. Pailin, "Abraham and Isaac: A Hermeneutical Problem before Kierkegaard," in *Kierkegaard's* Fear and Trembling, ed. Robert L. Perkins (Tuscaloosa: University of Alabama Press, 1981) 10–42.

[15] Witness the importance of James 1:17–22, "Every good gift and every perfect gift is from above," in his religious discourses.

Expectoration" (FT 34), there is no discussion of the divine nature, or even of the special qualities of Abraham's relationship to God that would render his obedience more intelligible morally. Instead, all attention is given to the horrific command itself and to the definitive way in which it leads Abraham outside any conceivable realm of ethical justifiability. If *Fear and Trembling* defends a divine command ethic, therefore, it is a forbidding and frightening ethic, indeed. Johannes's use of Genesis 22 suggests that the god of *Fear and Trembling* and his loyal devotee, Abraham, are more "beyond good and evil" than most commentators have wanted to admit.[16]

Read as an ethical treatise, *Fear and Trembling* leaves us strangely disturbed. Once we put aside the compulsion to ethicalize Abraham's conduct in the ways that violate the clear sense of the text, we are left with book whose exemplar borders on the psychopathic. Of course, this may be part of Kierkegaard's purpose in troubling his contemporaries' religious complacency. In that case, without really intending to offer Abraham in Genesis 22 as a model of behavior, Kierkegaard/Johannes would deliberately use this provocative and troubling episode to reinforce the book's call to personal religious engagement and commitment.

Before we conclude that *Fear and Trembling*'s treatment of ethics exists only for its shock value, it is worth considering whether it may point to another, and still deeper level of meaning in the text. This would be in keeping with Kierkegaard's program

[16] Philip Quinn has argued that the biblical Abraham faces a genuine and tragic moral dilemma: a conflict between two equally indefeasible requirements from which the agent "cannot escape wrongdoing and guilt."—"Agamemnon and Abraham: The Tragic Dilemma of Kierkegaard's Knight of Faith," *Journal of Literature & Theology* 4, no. 2 (July 1990): 183. See also Quinn's "Moral Obligation, Religious Demand, and Practical Conflict," in *Rationality, Religious Belief, and Moral Commitment*, ed. Robert Audi and William J. Wainwright (Ithaca, New York: Cornell University Press, 1986) 195–212.

of writing a book in such a way that it forces the reader to probe beneath its surface utterances. The presence of this still deeper level will have to be shown on the basis of textual evidence despite the author's attempt to conceal it. This further level might help solve the puzzle of why Johannes repeatedly commends a figure so dramatically "beyond ethics" as Abraham.

Level Four: Sin and Forgiveness

I believe this further level of meaning exists. At this level, we can read *Fear and Trembling* as addressing an abiding question of Christian faith: how can the individual believer be saved from sin? At this level, *Fear and Trembling* involves an exploration of each individual's inevitable encounter with the problems of moral self-condemnation and sin and the possibility of God's overcoming these through an act of divine grace. From the perspective of this level, all the other levels of *Fear and Trembling*—the focus on Abraham, the investigation of the psychology of faith, and the lengthy discussion of the "teleological suspension of the ethical"—have as their latent meaning the themes of sin and forgiveness.[17]

Questions related to soteriology are obviously important ones for Kierkegaard. The religious discourses that bracket *Fear and Trembling* in his authorship address them, and they are central themes in adjacent pseudonymous works like *The Concept of Anxiety, Philosophical Fragments,* and *Postscript.* Ordinarily, however, issues of sin and salvation are not seen to be a major preoccupation of *Fear and Trembling* itself. Its central figure, after

---

[17] Within the interpretive tradition, this understanding of *Fear and Trembling* has been offered by Louis Mackey in "The View from Pisgah: A Reading of *Fear and Trembling*," *Kierkegaard: A Collection of Critical Essays,* ed. Josiah Thompson (New York: Doubleday, 1972) 394–428 and Gregor Malantschuk, *Kierkegaard's Thought* (Princeton: Princeton University Press, 1971) 238.

all, is Abraham, and, Abraham, as Johannes repeatedly tells, "was a righteous man" (FT 99).

To perceive the importance of the themes of sin and grace in this book, we must invoke three different areas relevant to the text: (1) the tradition of interpretive commentary on Genesis 22; (2) the text of *Fear and Trembling* itself; and (3) aspects of Kierkegaard's biography.

The Interpretive Tradition

As David Lerch has shown, a long tradition of commentary exists that had already appropriated Genesis 22 for Christological purposes.[18] This tradition has its start in Galatians (3:13–14) with Paul's identification of Isaac with Christ as the "child of promise." It is picked up in Hebrews 11:17–19, whose author, presumably drawing on Jewish sources that held Isaac actually to have been sacrificed,[19] alludes to the Genesis episode as proof of the resurrection of the dead. Among the early Church fathers, these scriptural beginnings lead to a standard view of Abraham as a "type" or "figure" of God whose willing sacrifice of his son symbolizes God's involvement in the crucifixion.

We know that Kierkegaard was familiar with this tradition well before writing *Fear and Trembling* since he refers to it in a journal entry for 1839 (PAP, II A 569 13 September 1839 [JP1–298]). Equally important, this tradition was also familiar to Kant, whose treatment of Abraham in his *Religion within the Limits of Reason Alone* and *The Conflict of the Faculties* may have provided the stimulus for *Fear and Trembling*. Among other things, Kant's aim in these treatises is to deny the value of relying on historically mediated salvation as a solution to the problem of sin. Any such reliance, Kant maintains, threatens to usurp the place of rational

[18] David Lerch, *Isaaks Opferung christlich gedeutet* (Tübingen: J.C.B. Mohr, 1950).

[19] Shalom Spiegel, *The Last Trial* (New York: Pantheon Books, 1967).

conscience and moral striving in our redemption and to degenerate into immoral "superstition" (CF 119n.). In keeping with this theme, Kant offers Abraham as a negative example of someone who placed alleged divine commands above the clear dictates of rational conscience. Lest the reader miss the link between this criticism of Abraham and orthodox Christian ideas of salvation, Kant refers in a footnote of the *Conflict* to the tradition of viewing Abraham's willingness to offer his son as "a symbol of the world-savior's own sacrifice" (CF 121n.).

There is considerable evidence that Kierkegaard's specific defense of the importance of a historical savior in the *Fragments* and *Postscript* was a response to Kant's position in both these writings.[20] If so, we can regard *Fear and Trembling* as an opening salvo in this intense battle with Kant—and by extension with any rationalist philosophy (including Hegel's) that underestimates the seriousness of sin and the radical measures needed to overcome it. To introduce these issues, Kierkegaard employs the figure of Abraham in a limited and purely symbolic way. Abraham is not a sinner. *Fear and Trembling* offers none of the analysis of sin and its psychology found in works like *The Concept of Anxiety* or *The Sickness unto Death*. Instead, the text uses Genesis 22 to establish the possibility of a realm of faith above the realm of rational morality. Through Abraham's experience on Mount Moriah, we learn that God can transcend the ethical and can enter directly into the lives of those who themselves have transcended the ethical (for whatever reason). Without God, we are told Abraham is "lost" (FT 81). With God, Abraham the murderer becomes Abraham the father of faith. What Kierkegaard has done, in other words, is establish an analogy between Abraham and God, on the one hand, and between God and the rest of us on the other. Some features of this analogy (the possibility of an absolute relationship

[20] See my, *Kierkegaard and Kant: The Hidden Debt* (Albany: State University of New York Press, 1992). See also chapter five.

to God that suspends the ethical) are relevant to both sides of the analogy while others (the precise way in which we and Abraham suspend the ethical) are not.

This use of a biblical figure in a purely symbolic and typological way is not unique to *Fear and Trembling*. It also occurs in *Repetition*, with which *Fear and Trembling* was simultaneously published. Here it is Job who is used to explore the possibility of loss beyond "every *thinkable* human certainty and probability" of recovery (Rep 212). Like Abraham, Job gets everything back, including the family he had lost. Both men are undeniably innocent—Abraham because his deed results from a divine command, and Job by virtue of information given us in the prologue to the book. Yet Kierkegaard provides clues suggesting that each figure is a "type" for the loss and recovery represented by sin and forgiveness. In comments in his papers and in a draft for the manuscript of *Fear and Trembling* Kierkegaard considers presenting Abraham's previous life "as not devoid of guilt," with the result that the patriarch is led to "perceive the divine command as God's punishment" (PAP IV A 77 n.d., 1843 [JP5 5641]. Cf. PAP IV B 66 n.d., 1843). In *Repetition*, Job's righteousness is said to include "being proved to be in the wrong *before* God" (FT 212). This important theme is familiar from the "Ultimatum" of *Either/Or*, where it introduces the idea of sin.[21] Thus, in both *Fear and Trembling* and *Repetition* the suffering and redemption of innocent biblical prototypes is used to hint at the experience of ordinary mortals. If God's commands can imperil such paragons of virtue, what must these commands do to people like you and me caught up in frailty and sin? And if a personal relationship with God can redeem these men—returning to each the

[21] In the Supplement to *Repetition*, the Hongs' footnote at this point adds: "In a copy of *Enten/Eller*, II (*SV* II 306), Kierkegaard wrote: 'If a person is most fully in the right, before God he ought always have an even higher expression: that he is in the wrong, for no human being can penetrate his consciousness absolutely' (PAP IV A 256)."

descendants he appeared to have lost—what wonders can God's grace do in our lives? Shortly we will see that the familial aspect of the Abraham and Job narratives adds yet another dimension to the complex analogy Kierkegaard is constructing.

Themes of Sin and Grace

It is common to deny that sin forms much of a theme in *Fear and Trembling*.[22] Certainly, psychological and ethical issues capture most of the attention. Nevertheless, once we entertain the possibility that sin and forgiveness form an important, deeper level of meaning in the text, these themes become far more apparent. The book's title, for example, hearkens back to Paul's discussion of sin, grace, and redemption in Philippians 2:12 with its reminder that "God is at work in you, both to will and to work for his good pleasure." The Exordium obliquely refers to God's creation of free beings and their painful separation from him (FT 13), and there are other references to sin throughout the text.

Although sin and forgiveness are only touched on early in the book, they suddenly spring up before us in the story of Agnes and the merman that dominates Problema III. Examining the painful choices facing the merman who has seduced and then fallen in love with an innocent young woman, Johannes now embarks on an extended discussion of the problems of sin and repentance. "When the single individual by his guilt has come outside the universal," Johannes tells us, "he can return only by virtue of having come as the single individual into an absolute relation to the absolute" (FT 98).

The discussion continues in the text and a footnote with a critique of Hegel's concept of sin, with an observation on how sin takes the individual demonically higher than the universal, and with a statement that an ethics that ignores sin is a futile discipline

---

[22] Gene Outka, "God as the Subject of Unique Veneration: A Response to Ronald M. Green," *Journal of Religious Ethics* 21, no. 2 (Fall 1993): 211–215.

while one that acknowledges it has exceeded itself. These observations are all prefaced by the remark, "Now here I would like to make a comment that says more than has been said at any point previously" (FT 98), and they close abruptly with the reminder that "nothing of what has been said here explains Abraham, for Abraham did not become the single individual by way of sin" (FT 99).

It is no accident that Kierkegaard's choice of words in this discussion, especially his remarks about the repentant individual having to come "as the single individual into an absolute relation to the absolute" are identical with those Johannes uses earlier to describe Abraham's movement of faith. What Kierkegaard is letting us know here, is that Abraham and the merman are counterparts, positive and negative expressions of the same problem. Both have suspended the ethical, one by obedience and one by sin, and both are saved only by a direct, supraethical relationship to God. Once we understand that Abraham functions as a figure for the problem of sin and atonement, I think we also can see that this discussion of sin is not a chance aside but a window into *Fear and Trembling*'s deepest concerns.

Repetition provides confirming insight into Kierkegaard's authorial strategy in both these works. Like *Fear and Trembling*, most of *Repetition* ignores the problem of sin as it develops its special concern, the possibility of repetition in life. The focus is on the psychology of repetition and the experience of the young man who wishes to recover his alienated love. Job is introduced to expand the idea of loss and recovery, and near the end of this discussion, we find brief mention of Job as sinner (Rep 212). Yet in his Papers, Kierkegaard repeatedly informs us that the true repetition involves a return to the integrity lost by sin. In a lengthy unpublished reply to P.A. Heiberg's review of the book, he dismisses Heiberg's vague appeal to repetition as involving "spiritual development of a self-conscious free will." Remarking that repetition "cannot be left in this nebulous way," he tells us that "it is a question of nullifying the repetition in which evil recurs and of bringing forth the repetition in which good recurs" (PAP IV B 111 n.d.. 1843–44, p. 267). In its highest sense, he states, repetition is atonement.[23] In a long entry signed by the pseudonym Constantin Constantius, we are offered insight into the deeper intention of the book:

> *Repetition* was insignificant, without any philosophical pretension, a droll little book, dashed off as an oddity, and curiously enough, written in such a way that, if possible, the heretics would not be able to understand it.... [T]he true repetition is eternity; however, that repetition (by being psychologically pursued so far that it vanishes for psychology as transcendent, as a religious movement by virtue of the absurd, which commences when a person has come to the border of the wondrous), as soon as the issue is posed dogmatically will come to mean atonement...." (PAP IV B 120 n.d., 1843)[24]

The themes in this comment should by now be familiar to us from *Fear and Trembling*: a text with a deliberately hidden message; the movement from psychology to dogmatics; and the intensification of religious consciousness "by virtue of the absurd." The considerable parallelism between these two works provides confirming evidence that themes of sin and grace are far more salient at this point in Kierkegaard's authorship than they are commonly assumed to be.

Biographical Correspondences

The ordinary cautions against using the facts of an author's life to interpret his writings have special relevance to a writer like

---

[23] This remark is in a preliminary draft of PAP IV B 117 that appears in PAP IV B 118 n.d., 1843–44, p. 300. This passage is translated in the "Supplement" to Rep 320.

[24] This passage is translated in the "Supplement" to *Repetition* (FT 324).

Kierkegaard. As the Hongs observe, "no writer has so painstakingly tried to preclude his readers' collapsing writer and works together and thereby transmogrifying the works into autobiography or memoir" (FT xi). Despite this, there can be no doubt that *Fear and Trembling* is among the most personal of Kierkegaard's writings. The events of his broken engagement to Regine Olsen provided an immediate stimulus for the book, and tales of frustrated love and marriages blocked by fate abound in it.[25] In his papers Kierkegaard states boldly, "He who has explained this riddle has explained my life" (PAP IV A 76 n.d., 1843 [JP5 5640]).

Most commonly, the biographical correspondences here are linked to the ethical themes of the text. In his selection of Genesis 22, Kierkegaard is seen as providing Regine—his secret reader—an explanation of their broken relationship. Just as Abraham received a divine command to sacrifice what was dearest in his life, so Kierkegaard was compelled to obey the divine "governance" and set his aside his worldly hopes of happiness in order to undertake his solitary vocation as a religious author. Regine herself contributed to this reading in later statements that cast the breach in these terms.[26] This focus on conflicting ethical responsibilities and priorities also fits nicely with the view that the Problemas are primarily a rejection of Hegelian *Sittlichkeit* with its preference for social role responsibilities over individual existence.

---

[25] These include mention of the young swain and princess (FT 41–45); a young girl forced by her parents to marry someone other than the one she loves (85); a young swain blocked from possessing his beloved because doing so will "destroy a whole family" (85); a bridegroom "to whom the augurs prophesied a calamity that would have its origin in his marriage" (89–92); Queen Elizabeth's sacrifice of her love, Essex, for the state (93–94); the tale of Agnes and the merman (94–99); and the story of Sarah and Tobias (102–106).

[26] Steen Johansen, *Erindringer om Søren Kierkegaard* (København: C.A. Reitzels Boghandel. 1980,) 32–44.

In all his discussions of the biographical events that underlay the book, however, Kierkegaard offers a very different explanation. God, he tells us, had issued a "no" to the marriage. The reasons for this lay not in any call to a religious vocation but in Kierkegaard's own melancholy and sense of perdition—what he called "the eternal night brooding within me" (PAP IV A 107, 17 May 1843, p. 43 [JP6–5664]). This melancholy was itself the fruit of a familial tradition of sin begun by his father, Michael Pedersen Kierkegaard. The elder Kierkegaard's youthful curse of God from a hillock on Jutland heath and his sexual sins, including his extramarital relationship with Kierkegaard's mother, the handmaid Ane Lund, following his first wife's death,[27] had led him to see himself as fated to lose his children as a punishment. Indeed, he did live to bury five of the seven Kierkegaard children. If there is a link between the Abraham story and Kierkegaard's life, therefore, it is the peril to which the elder Kierkegaard's acts had exposed the family. In this context, it is not Regine who plays the role of Isaac but Søren himself.[28]

Abundant textual evidence supports the claim that the father's conduct forms a major biographical substratum to the book. Father-child themes abound, from the epigraph's opening mention of Tarquinius Superbus's secret message to his son, through the numerous tragic heroes whose conduct imperils their offspring, to the choice of Abraham. (In *Repetition* it is again a father, Job, whose relationship to God threatens his family). The book's several tales of frustrated marriages—including Aristotle's story of the young man whose marriage threatens to "destroy a whole family"; and the stories of Sarah and Tobias and Agnes and the merman—all either assume family lines bearing a curse or

[27] Sylviane Agacinski argues that the unmentioned sin that underlay the elder Kierkegaard's sense of perdition was the rape of the handmaid, Ane Lund. See her *Aparté: Conceptions and Deaths of Søren Kierkegaard*, trans. Kevin Newmark (Tallahassee: Florida State University Press, 1988) 244ff.

[28] Malantschuk, *Kierkegaard's Thought*, 238.

unions blocked by a sinful past. *Fear and Trembling* may be a message to Regine, but it is not simply about Kierkegaard's call to a religious vocation. Rather, it is an explanation of why Søren, himself both sacrificer and sacrificial victim, Abraham and Isaac, had acted to spare her involvement in his family's melancholy fate.

Not that *Fear and Trembling* is wholly negative in this regard. Read as an examination of faith and grace, it's larger message, whether directed to Kierkegaard's deceased father, to Regine, or to Kierkegaard himself, is one of hope. God's command imposes on Abraham a grueling ordeal, but in the end, God's sovereignty over ethics triumphs. Not only can God command murder, but he can make a murderer the father of faith. As Abraham discovered, God's last word is not death and condemnation. God can effect a teleological suspension of his justice to renew his relationship with an individual. With this anguished religious hope in mind, we can also perhaps better understand Johannes's repeated expressions of doubt that he could exhibit the kind of faith shown by Abraham. A sense of inescapable sin and familial disaster had blocked Kierkegaard's marriage. Perhaps, in faith, one could make the absurd leap that Abraham did, believing that God could snatch life from the jaws of death and somehow continue a family line otherwise doomed to extinction. However, neither Kierkegaard nor his pseudonym is sure he possesses Abraham's faith. Against the background of these difficult beliefs we can better understand Kierkegaard's remark in his journal: "If I had had faith, I would have stayed with Regine" (PAP IV A 107, 17 May 1843, p. 41 [JP5–5664]).[29]

---

[29] See George Pattison, "A Drama of Love and Death: Michael Pederson Kierkegaard and Regine Olsen Revisited." *History of European Ideas* 12, no. 1 (1990): 79–91 for a discussion of the complexity of the relationship between Kierkegaard's biography and his writings and for a treatment of the intricate ties between Kierkegaard's relationship to his father and his relationship to Regine. For a more traditional view of Kierkegaard's remark as relating

Towards a Transparent Text

I have argued that as we proceed through *Fear and Trembling* we come across new themes and ever-deeper levels of meaning. Beginning as an impassioned call to lived Christianity, the text leads through discussions of the psychology of faith and the ethical outlines of the Christian life and finally to themes of salvation, grace, and forgiveness.

Kierkegaard's proposes a metaphor of the text as progressively disclosed by a caustic fluid. I want to close with the suggestion that a comprehensive reading of *Fear and Trembling* aims at a text that is a fully developed and transparent image. While each level of meaning preserves its independent significance, the cumulative meaning grows with each new level of disclosure until we arrive at a penetrating view where each level of meaning is superimposed on and enhances the others.

For example, the meaning of the call to lived commitment deepens as we encounter the demanding psychology of faith. Here the text not only challenges bourgeois complacency but uses a consummately bourgeois "knight of faith" to hint at the depths of suffering and interiority that mark true Christian faith. As we turn to the Problemas, both the call to commitment and the psychology of faith receive new significance from an outline of the normative demands of Christian life. With the suggestion that Christians may be required to go beyond the confines of family or nation to establish their own relation to the absolute, the themes of commitment, loss, infinite resignation, and faith are amplified and made concrete. Christian love emerges as selfless care rooted in the psychic renunciation of all proprietary claims.

Finally, we reach the level where themes of sin and grace predominate. Looked back on from here, the call to commitment is

---

primarily to his choice of a religious vocation over marriage to Regine, see Sylvia Fleming Crocker, "Sacrifice in Kierkegaard's *Fear and Trembling*," *Harvard Theological Review* 68 (April 1975): 125–139.

now seen to involve repentance: the awareness that one's life, however accomplished and successful by outward measures, stands under judgment. Applied to culture, this also becomes a critique of Hegelianized Christian civilization for its superficiality, pride, and obliviousness to sin. At the level of psychology, the two movements of faith also take on new meaning. Infinite resignation is now seen to require an abandonment of one's sense of moral integrity and an acknowledgment of the reality of sin (a movement the merman can make), whereas faith, the second movement, becomes an absurd hope of redemption and renewal beyond all ones reasoned claims or expectations (a movement beyond the merman's powers).

The awareness of sin and grace also permeates the specifically ethical level of the text with new significance. The radicalness of the Christian ethic—the possibility that one may be called to individual existence beyond family or state—sharpens obligations to the breaking point and eliminates any false sense of one's ability to comply with God's commands. Acceptance of Jesus' life as the pattern for one's own—including the command to "hate" one's father and mother (Luke 14:26)—establishes an ethic requiring virtually inhuman commitment. One who is aware of these ethical demands must conclude that, "Before God, we are always in the wrong." At the same time, an appreciation of the depth of even our mundane human sins, our greed, lusts, and anger, renders naive any merely social definition of the self. Hegelian *Sittlichkeit* runs aground on its own spiritual shallowness. Here *Fear and Trembling* tells us that relationship to God, as judge and redeemer, takes primacy—and must precede—any social integration of the self. Seen in this way, as the acknowledgment of sin and acceptance of grace, the teleological suspension of the ethical becomes just that: a suspension of ethics rather than its annulment. Grace aims at one's full moral renovation.[30] An awareness of God's gracious forgiveness ends self-obsession and pride, and elicits the selfless care that is morality's highest *télos*.

Earlier, we noted a deep tension at the ethical level of *Fear and Trembling* between the admiration repeatedly voiced for Abraham and the equally clear assertions that, ethically speaking, what he does amounts to murder. We saw that some commentators have tried to overcome this tension by inappropriately ethicalizing Abraham's conduct. Viewed in relation to the themes of sin and grace, however, these seemingly opposed aspects of *Fear and Trembling* can be seen to comfortably reside besides one another. Like Paul and Luther before him, Kierkegaard can celebrate Abraham's transcendence of the moral law—in the sense that nothing whatsoever in his ethical conduct warrants his election or renown—while simultaneously holding him up as a model for all to emulate. Precisely because he is justified by grace alone, Abraham is deservedly the "father of faith." He is also a beacon to all those who knowingly "suspend the ethical" in a frank admission of sin and look to God alone for their salvation.

Conclusion

Fear and Trembling has earned renown as a provocative statement of challenge. But it is more than that. *Fear and Trembling* contains the major themes of Christian faith and ethics that will emerge in the ensuing pseudonymous works and many of the religious discourses. *Fear and Trembling* deserves the fame that Kierkegaard predicted for it, but that very fame may have obscured the fact that this is no eccentric statement by a youthful poet. It is a profound theological treatise firmly rooted in the Pauline/Lutheran tradition to which Kierkegaard belonged.

---

[30] This conforms to the text's insistence that only a *suspension* of ethics is being implied—FT 54.

7

Fear and Trembling: A Jewish Appreciation

One does not think of Søren Kierkegaard as a friend of the Jews. Kierkegaard's published writings and journals are marred by anti-Semitic asides and sometimes repeat familiar anti-Semitic stereotypes. As Bruce H. Kirmmse has noted, Kierkegaard's hostility to Judaism as a religious tradition and as a mode of religious being may have intensified over the course of his life, becoming conflated with his growing enmity to the culture of "Christendom."[1] In both Judaism and the surrounding cultural Christianity, Kierkegaard perceived a stance of accommodation to worldly concerns that he believed was the opposite of the position called for by authentic New Testament faith.

Despite this undeniable theme in Kierkegaard's writings, I believe it is unfair to label Kierkegaard an anti-Semite. His many unfavorable allusions to Jews and Judaism lack any hint of the hatred and personal involvement of the committed anti-Semite. As Kirmmse shows, Kierkegaard's use of Jewish stereotypes has little to do with Jews and Judaism, "but is principally a part of Kierkegaard's battle against the lukewarm and flimsy Christendom of his times."[2] Most important, Kierkegaard's basic religious stance—his religious individualism and his unwavering call to personal choice and responsibility—is deeply opposed to

---

[1] Bruce H. Kirmmse, "Kierkegaard, Jews, and Judaism," *Kierkegaardiana* 17 (1994): 83–97.

[2] Ibid., 94.

the kind of "in-group/out-group" mentality that has always fueled anti-Semitism.

Another, less obvious, sign of Kierkegaard's resistance to anti-Semitic modes of thinking is his adoption of a theological stance that refuses to relinquish some of the leading ideals of biblical faith. Nowhere is this more evident than in the treatment of Abraham found in *Fear and Trembling*. By upholding Abraham as a "knight of faith," *Fear and Trembling* adopts what is essentially the classical position of both Hebrew and Jewish faith. It does so, moreover, in a cultural environment where the idea of Abraham as a model of faith was under sharp attack. Nearly two generations earlier, Kant had developed a rationalist critique of Abraham in two of his mature writings, *Religion within the Limits of Reason Alone* and *The Conflict of the Faculties*. Although not identifiably anti-Semitic, Kant's treatments of Abraham reflect Enlightenment discomfort with revealed religion in any of its forms and provide ammunition for left wing critics of many aspects of Jewish faith. In the generation immediately preceding Kierkegaard's, Hegel had developed a stinging critique of Judaism in general and Abraham as its leading representative. This critique both reflects and fed right wing, nationalist anti-Semitism. What is remarkable in this conceptually hostile environment is that *Fear and Trembling* eschews both these directions. In doing so, it tells us something about Kierkegaard's orientation toward the Hebrew and Jewish sources of Western faith. Even if we concede that the views of Johannes de Silentio do not represent Kierkegaard's in all respects, there is no reason to believe that on this one central theme of *Fear and Trembling*—Abraham as a religious exemplar—Kierkegaard or his pseudonym disagree.

In what follows I want briefly to compare the treatment of Abraham found in *Fear and Trembling* with three alternatives. The first is the broad tradition of Judaic reflection on Genesis 22. We will see that while there is no single Jewish view of this episode, the available alternatives share some very important common features with the general view of Abraham presented in *Fear and Trembling*. Next, I will sketch the Kantian critique of the Genesis episode. Kant's willingness to question the biblical account stands in marked contrast to the entirety of the Jewish tradition. Finally, I will review Hegel's viciously anti-Semitic treatment of Abraham. That *Fear and Trembling* evidences not a hint of the kinds of posture adopted by Hegel reinforces our understanding of Kierkegaard's opposition to the spirit of Hegelian philosophy; it also evidences the depth of Kierkegaard's opposition to the kinds of intellectual position that consistently underwrite anti-Semitism.

Jewish Perspectives on Genesis 22

In classical Jewish thought, the episode of Genesis 22 is called the *Akedah* or "binding" in reference to the ties needed to restrain Isaac as a sacrificial victim. Jewish writings on the *Akedah* stretch across centuries of commentaries, and include such diverse materials as rabbinic Midrashim and Philo's or Maimonides's widely separated efforts to harmonize Abraham's conduct with the traditions of Greek and Roman philosophy. Louis Jacobs argues that despite this diversity of materials, Jewish thinking can be arrayed in terms of three identifiable attitudes toward the Genesis episode.[3] The first stresses the story's "happy ending." What is emphasized here is the fact that, while God had various reasons for putting Abraham to the test, he never intended to permit the slaughter of Isaac. As Shalom Spiegel has shown, these ideas became extremely important to Jews during the frequent pogroms during the period of the Crusades.[4] It then became

[3] Louis Jacobs, "The Problem of the *Akedah* in Jewish Thought" *Kierkegaard's* Fear and Trembling: *Critical Appraisals,* ed. Robert L. Perkins (Tuscaloosa: University of Alabama Press, 1981) 1–9.

[4] Shalom Spiegel, *The Last Trial,* trans. Judah Goldin (New York: Schocken Books, 1950, 1969).

commonplace in Jewish piety to appeal to the *Akedah.* As God had spared Isaac, so he would spare Jewish children from death at the hands of enemies. Within this whole line of approach, one very important subtradition speculates that Isaac may actually have perished. His "return" to the living is thus advanced as proof of the resurrection of the dead. Whatever the specifics of the interpretation, the focus is not on the command but on the episode as proof of God's ultimate justice and redemptive intent.

A second Jewish attitude focuses on the command itself and Abraham's obedience. Little effort is made to rationalize the test or morally justify God's ways. Abraham is upheld as a model of faith because of his willingness to place obedience to God above any other obligation or motive he may have. In this tradition of Jewish thought, says Jacobs, the ultimate for Abraham "is not the ethical norm but his individual relationship to his God."[5]

A third attitude combines features of the first two. In this view, it is seen to be impossible in reality that God could ever be false to his nature or his promises and command Isaac's murder. Yet if God should command such a thing, Abraham, and all that follow in his path, must be prepared "to cross the fearful abyss."[6] Faith is the confidence that God's righteousness and God's commands can never ultimately diverge. This third attitude brings together Judaism's intense commitment to God's moral nature and its equally strong commitment to the model of radical obedience upheld in Genesis 22.

There is nothing in *Fear and Trembling* that resembles the first Jewish attitude. Johannes de Silentio makes no effort to provide an ethical justification of God's command or Abraham's obedience. However, positions close to the second and the third are both

---

[5] Louis Jacobs, "The Problem of the *Akedah* in Jewish Thought," in *Kierkegaard's Fear and Trembling: Critical Appraisals,* ed. Robert L. Perkins (Tuscaloosa: University of Alabama Press, 1983) 2.

[6] Ibid.

found in the text. The second attitude finds support in Johannes's repeated disavowals of any possible moral interpretation of Abraham's conduct and of the divine motivation that elicited it. Throughout, he emphasizes the stark command and Abraham's obedience. The Genesis episode is held to involve a "teleological suspension of the ethical" in which the father's moral duty to care for the son is subordinated to his individual relationship to God. A "tragic hero" like Agamemnon also offers up a child to appease the gods, but his goal of ensuring victory for his nation in war is ultimately understandable in human and ethical terms. In contrast, says Johannes, "Abraham is at no time a tragic hero but is something entirely different, either a murderer or a man of faith" (FT 57).

Although *Fear and Trembling* can be read to support this radically non-moral understanding of the divine command and religious faith, other, equally well-grounded readings bring it closer to the third Jewish attitude signaled by Jacobs, one that sees Abraham's behavior as joining a willingness to obey with unwavering confidence in the ultimate goodness of God's purposes. Gene Outka, for example, reminds us that we must read this book against the background of Kierkegaard's belief that "God is love," a belief that is reaffirmed at least once early in the text by Johannes de Silentio (FT 34). Because of this belief, says Outka, we should understand Abraham as combining "general trust" in God's goodness with "specific perplexity" about the meaning of the command.[7] Abraham's faith is not a willingness to commit murder at God's command, but the seemingly absurd confidence that in all circumstances and whatever the appearance, God's love prevails. This idea of faith, not as irrational immoral obedience but as the trust in God's ultimate righteousness, is also

---

[7] Gene Outka, "Religious and Moral Duty: Notes on *Fear and Trembling*," in *Religion and Morality*, ed. Gene Outka and John Reeder, Jr. (Garden City, New York: Anchor Books, 1973) 212.

deeply Judaic. It has been a sustaining feature of Jewish life over centuries of hardship and persecution.

Before leaving *Fear and Trembling,* I would add that there is still another Jewish understanding of Genesis 22—a fourth Jewish attitude beyond those mentioned by Jacobs—that also has a remarkable presence this text. This sees the *Akedah* as proof of God's justice-transcending mercy and his willingness to forgive sin. In the Jewish sources, this understanding has several bases. One is the view that by their obedience, Abraham and Isaac established a reserve of merit. Father and son become a vicarious sacrifice for future generations. In rabbinic thought, this idea is furthered by the linking of the *Akedah* with the temple sacrificial cult. For later piety and popular tradition, virtually all the elements of the Genesis episode assume a cultic significance. Moriah is said to be the site of the temple mount, the ashes of the sacrificial fire (Isaac's ashes in some accounts) underlie the altar and the various parts of the ram are identified with cultic objects, its horn becoming converted into the shofar blown on Rosh Hashanah, the Jewish New Year's Day, in anticipation of the season of atonement.[8]

Against this symbolic and conceptual background, certain rabbinic appeals to the *Akedah* make sense. One Midrash, for example, expands on the meaning of Genesis 22:13, "after him caught in the thicket," by relating it to the future:

> The Holy One, blessed be He, meant: Behold what is to come! Thy children who will succeed thee will one day be entangled and caught in sins like the ram in the thicket. What use are they then to make of a ram's horns? They are to lift up the horns and blow

[8] Ronald M. Green, *Religion and Moral Reason: A New Method for Comparative Study* (Oxford University Press, 1988) 99.

> them. Whereupon I will be reminded of the binding of Isaac and will acquit them in the judgment.[9]

Elsewhere, in a similar vein, the thicket is associated with the many foreign nations with whom the Israelites' sins will involve them, and the *Akedah* is appealed to as the sign and promise of redemption.

Another version of the *Akedah* as an appeal to God's mercy draws on Abraham's willingness to subordinate his own immediate emotional needs to the divine command. A text of the Sephardic Rosh Hashanah liturgy goes so far as to ask God to model his own behavior on Abraham's:

> And may there appear before Thee the *Akedah* to which Father Abraham subjected his son Isaac on the altar, suppressing his own feelings of pity in order to do Thy will with a perfect heart. So let Thy feelings of pity suppress Thine anger (and remove it) from us.... May Thy compassion suppress Thine anger, oh, may Thy compassion prevail over Thine (other) Attributes.[10]

In several articles and chapters published over the past fifteen years I have maintained that it is mistaken to read *Fear and Trembling* primarily as a work focused on ethical questions concerning the norms of human conduct.[11] Instead, with Louis Mackey and others, I have argued that, at a much deeper but pervasive level, the book is a discussion of Christian soteriology. Like Christian commentators before him, Kierkegaard uses the Abraham story to develop the classical Pauline doctrine of

---

[9] *Pesikta Rabbati,* trans. William G. Braude (New Haven, Yale University Press, 1968) 21.

[10] Shalom Spiegel, *The Last Trial,* 89.

[11] Ronald M. Green, "Deciphering Fear and Trembling's Secret Message," *Religious Studies* 22 (1986): 95–111. "Enough Is Enough! *Fear and Trembling* is *Not* about Ethics" and "A Reply to Gene Outka" Journal of Religious Ethics 21, no. 2 (Fall 1993): 217–220; *Kierkegaard and Kant: The Hidden Debt,* chapter five; "'Developing' *Fear and Trembling*" (reprinted here as chapter six).

justification by faith through grace. The "teleological suspension of the ethical" imagined by the text, therefore, is not primarily a suspension of reasoned moral norms in the name of obedience to God. Rather, it expresses God's own ability to put aside his just punishments of human sinfulness to establish an absolute relationship with each individual beyond the demands of justice.

This, of course, is the same theme found in this fourth strand of rabbinic speculation. Its presence in Kierkegaard's thinking should not be surprising. The soteriological use of the Genesis episode is present in both early rabbinic and in Pauline sources, where it is turned to explicitly Christian purposes. Kierkegaard thus stands at the end of a long common tradition. At the close of my remarks, I will return to the question of what this commonality means for our understanding of Kierkegaard's relationship to Judaism. For now, however, what is important to note is that *Fear and Trembling*, for all its rootage in the philosophical debates of their early nineteenth century, is also a classical text of biblical theology. Indeed, if the few open Christian references were removed, it could even be mistaken for a traditional rabbinic Midrash on Genesis 22.

Kant's Treatment of the Genesis Episode

Kant's treatments of Genesis 22 in the *Religion* and *The Conflict of the Faculties* represent a watershed in the long tradition of commentary on Genesis 22.[12] For perhaps for the first time in history, a philosophical or theological commentator feels free to call the patriarch's conduct under review, and with that, implicitly to criticize the religious tradition in which he stands.

The criticism in the *Religion* is rather muted and indirect. One comment arises during Kant's discussion of an "inquisitor" who, on the basis of the commands of "a supernaturally revealed

[12] David Lerch, *Isaaks Opferung christlich gedeutet*; Carol Delaney, *Abraham on Trial* (Princeton: Princeton University Press, 1998).

Divine Will," insists on persecuting heretics. In Kant's view, this inquisitor is caught in a conflict between ordinary human duty, which prohibits killing persons because of their religious beliefs, and obedience to the divine command. In a concession that is somewhat odd given his view that our experiential knowledge is necessarily limited to what is given by our senses, Kant allows that an explicit command from God "made known in extraordinary fashion" might override human duty in such cases. But, he quickly adds, "that God has ever uttered this terrible injunction can be asserted only on the basis of historical documents and is never apodictically certain." He then continues,

> ... and even did it appear to have come to him from God himself (like the command delivered to Abraham to slaughter his own son like a sheep) it is at least possible that in this instance a mistake has prevailed. But if this is so, the inquisitor would risk doing what would be wrong in the highest degree; and in this very act he is behaving unconscientiously. This is the case with respect to all historical and visionary faith; that is, the *possibility* ever remains that an error may be discovered in it. Hence it is unconscientious to follow such a faith with the possibility that perhaps what it commands or permits may be wrong, *i.e.*, with the danger of disobedience to a human duty which is certain in and of itself. (Rel 175)

Earlier in the *Religion*, Kant offers a more oblique, but in some ways sharper criticism. During a discussion of miracles, he remarks: "... even though something is represented as commanded by God, through a direct manifestation of Him, yet, if it flatly contradicts morality, it cannot, despite all appearances, be of God (for example, were a father ordered to kill his son who is, so far as he knows, perfectly innocent)" (Rel 81f.).

In the somewhat later *Conflict of the Faculties*, Kant's pulls together his epistemologically based rejection of our ability to "hear" God's voice and his defense of the priority of reasoned duty over any supernatural revelations to articulate an explicit denunciation of Abraham's conduct. In place of obedience to the murderous divine command, Kant tells, us

> Abraham should have replied to this supposedly divine voice: "That I ought not to kill my good son is quite certain. But that you, this apparition, are God—of that I am not certain, and never can be, not even if this voice rings down to me from (visible) heaven. (CF 115)[13]

Kant's ethical intensity and his willingness to place moral duties over religious obligations are not altogether remote from the Jewish tradition. This stance, after all, is inherent in prophetic faith, and it even finds expression in the Abraham narratives, such as the Sodom and Gomorrah episode, where the patriarch is presented as interrogating God and reminding him of his own lofty standard of justice.[14] The Kantian treatment of Genesis 22 is, thus, in some ways a natural outgrowth of biblical—and Judaic—faith, even if it will eventually make possible the modern break with revealed religion. At the same time, however, the explicit criticism of Abraham has no presence at all in the classical Jewish tradition. There, Abraham is seen as a virtual embodiment of the spirit of Jewish faith.

There is little doubt in my mind that *Fear and Trembling* is meant, among other things, to be a deliberate reply to Kant's criticisms of Abraham and Abrahamic faith.[15] This does not make Kierkegaard a philo-Semite. Abraham, after all, had long ago been appropriated by Christianity as the archetypical believer justified by faith alone. The defense of Abraham is, therefore, primarily a

---

[13] On page 115, there is a small typographical error in Gregor's translation that I have corrected.

[14] Emmanuel Levinas reminds us of this point and asks why Kierkegaard neglects it in his rendering of Abraham's conduct. See his *Proper Names* (Stanford, California: Stanford University Press, 1996) 74, 77.

[15] See my *Kierkegaard and Kant*, chapter five.

reply to the Kantian challenges to traditional Christian faith. Nevertheless, that *Fear and Trembling* chooses to use the patriarch, and not a New Testament figure, evidences Kierkegaard's loyalty to the shared Jewish and Christian tradition.

Hegel on Abraham

This loyalty becomes particularly striking when we compare it with Hegel's treatment of the patriarch in his *Early Theological Writings*.[16] Hegel deals with Abraham and the Genesis 22 episode in a long essay within this collection entitled "The Spirit of Christianity and Its Fate." The aim of this essay, among other things, is to vindicate Christianity as a religion of love and beauty over against the Judaism from which it emerged. There is no reason to believe that Kierkegaard read Hegel's *teologische Jugendschriften*, which were first published (in German) only in 1907. Nevertheless, the spirit of this early treatment of Abraham and Judaism is recapitulated in Hegel's later *Lectures on the Philosophy of Religion*, with which Kierkegaard was familiar.[17]

Hegel begins by identifying Abraham as "the true progenitor of the Jews," whose "spirit is the unity, the soul, regulating the entire fate of his posterity."[18] By studying Abraham, therefore, we can better characterize and understand Judaism and the Jewish people.

According to Hegel, the story of Abraham's life begins with the act by which, without having been either injured or disowned,

---

[16] Georg Wilhelm Friedrich Hegel, *Early Theological Writings*, trans. T.M. Knox (Chicago: University of Chicago Press, 1948).

[17] Niels Thulstrup, "Kierkegaard and Hegel," in *Kierkegaard and Speculative Idealism*, ed. Niels Thulstrup (Copenhagen: C.A. Reitzels Boghandel, 1979) 111. Hegel's treatment of Judaism in the *Lectures on the Philosophy of Religion* does not focus on Abraham. Nevertheless, many of the themes found in the *Jugendschripten* are recapitulated here, especially those sections dealing with the nationalism of the Jewish people, vol. 2, 193–219.

[18] Hegel, *Early Theological Writings* 182.

he tears himself free from his family and homeland. By means of a severance that "snaps the bonds of communal life and love," Abraham seeks to be "an overlord," "a wholly self-subsistent, independent man." Greek heroes like Cadmus or Danaus had also forsaken their fatherland, "but they forsook it in battle; they went in quest of a soil where they would be free and they sought it that they might love." By "their gentle arts and manners" these heroes "won over the civilized aborigines and intermingled with them to form a happy and gregarious people." Abraham, in contrast "wanted not to love, wanted to be free by not loving."[19] Even his nomadism was a deliberate disavowal of relationships with men or nature. He chose to wander "hither and thither over a boundless territory without bringing parts of it any nearer to him by cultivating and improving them." In his travels he avoided all human ties. Always, he persisted in cutting himself off from others, and even made this aim conspicuous by the "physical peculiarity" (circumcision) that he imposed on himself and his posterity. "When surrounded by mightier people as in Egypt and Gerar" or "in dealing with kings who intended no evil," Hegel continues, "he was suspicious and resorted to cunning and duplicities." But, whenever he was the stronger, as in opposing the five kings, "he fell about him with the sword."[20]

The crescendo of Abraham's detachment and even hostility to the world is reached for Hegel in the Genesis 22 episode. Speaking of this, he states:

> Love alone was beyond his power; even the one love he had, his love for his son, even his hope ot posterity—the one mode of extending his being, the one mode of immortality he knew and hoped for—could depress him, trouble his all-exclusive heart and disquiet it to such an extent that even this love he once wished to destroy; and his heart was quieted only through the certainty of

[19] Ibid., 185.

[20] Ibid, 186.

> the feeling that this love was not so strong as to render him unable to slay his beloved son with his own hand.[21]

The contrast here with *Fear and Trembling*, where Abraham's love for his son is a condition of the faithfulness of his obedience, could not be more complete.

Hegel's essay continues in this vein, bodying forth one malignant stereotype after the other of Abraham as "the eternal Jew." In short order, the diatribe moves on to other patriarchs and to Moses and the Exodus. Hegel presents this, not as a celebration of human freedom, but as a servile and cowardly flight by a people caught between feared masters. Throughout, Hegel tells us, the spirit of this people is "universal enmity" in which room is left for nothing "save physical dependence, an animal existence which can be assured only at the expense of all other existence, and which the Jews took as their fief."[22]

It is important to note that these vicious stereotypes are not merely given scholarly prestige by Hegel's reiteration of them. They are also woven into a philosophical tapestry that relegates Jews and Judaism to an ugly and primitive human past. The distinguished modern Jewish philosopher, Emmanuel Levinas, reviews the treatment of Abraham in the context of Hegel's philosophy.[23] Levinas recognizes that, among other things, Hegel is using Judaism to build toward a more embracing critique of nationalism. Nevertheless, he summarizes the baleful impact of Hegel's essay when he says,

> As for Judaism, the critical discourse is translated ... by a whole doctrine that corroborates (is it its source, or is it, despite all Hegel's greatness, its consequence?) the argument that, up to the present day, has nurtured anti-semitism... [F]rom this point on,

---

[21] Ibid., 187.

[22] Ibid., 191.

[23] Emmanuel Levinas, "Hegel and the Jews," in *Difficult Freedom: Essays on Judaism* (Baltimore: The Johns Hopkins University Press, 1990) 235–38.

> we get a virulent formula in which the enemy of the Jew will neither bother to understand nor, above all make understood the ambiguity of terms. Anti-Semitism is based within the System, which amounts to saying within the Absolute. What a godsend![24]

Hegel's treatment provides a solid indication of the extent to which the shared Jewish and Christian heritage, so powerfully represented by the figure of Abraham, was under assault in the early nineteenth century, even by some who would try to defend Christian faith. Against this background, what is noteworthy is Kierkegaard's refusal to pick up any of these culturally available themes. Abraham remains the epitome of faith, not the progenitor of a despised people or a morally surpassed religion.

Conclusion

I want to be very reserved in my conclusions. First of all, I certainly do not want to say that Kierkegaard, by choosing to utilize the Genesis 22 episode as the basis of *Fear and Trembling* sought to strike a blow for Jews or Jewish faith against Hegel or, more broadly against anti-Semitism. Although *Fear and Trembling* clearly takes issue with Hegelian philosophy and even with the reduction of human obligation to the confines of any human communities, it does so within a general defense of biblical faith, not of Judaism.

Second, it is important to recognize that the appeal to Abraham by a Christian thinker can hardly be taken as philo-Semitic, since this appeal itself belongs to a long tradition of Christian polemics against Judaism. For Paul, Abraham represented faith prior to the existence of the law; he is the progenitor, not of those who descend from him in the flesh (read: Jews) but of those who serve him in the spirit (the members of the new Christian community). If we add to this a long tradition that

[24] Ibid., 236.

sees Genesis 22 as a "figure" for the supreme father-son sacrifice, the crucifixion, and recognize that *Fear and Trembling* presumes this interpretive tradition,[25] then the suggestion of philo-Semitism becomes even less tenable.

My Dartmouth colleague, Susannah Heschel, has suggested that it is time to bring the categories of post-colonial theory to the understanding of Jewish and Christian relations.[26] In Heschel's view, from the beginnings of the Common Era, Judaism underwent a process of colonization that extended beyond territorial expropriations and political control to include aggressive cultural imperialism. Christian writers, she notes, invaded Jewish holy books, rituals, and sites, claimed them as their own, and placed the original inhabitants in a subservient position within their own traditions.

Christian uses of the Genesis account do not escape this harsh criticism. Neither does *Fear and Trembling*. To the extent that it seeks to use the Genesis episode to defend a Christian doctrine of justification by faith through grace, it stands within this long tradition of appropriation of the *Akedah* for Christian theological purposes. Nevertheless, when we consider the increasingly anti-Jewish interpretive models that were available to a Christian writer living and working within the context of northern Europe in the mid-nineteenth century, *Fear and Trembling*, in its very traditionality, remains a happy exception.

At the close of his essay on Hegel's *Early Theological Writings*, Levinas asks how we should respond to a philosophical position

---

[25] Green, *Kierkegaard and Kant*, 205.

[26] Susannah Heschel, "Revolt of the Colonized: Abraham Geiger's Wissenschaft des Judentums as a Challenge to Christian Hegemony in the Academy," *New German Critique* 77 (Spring/Summer 1999): 61–86. Also, "Theology as a Vision for Colonialism: From Supersessionism to Dejudaization in German Protestantism," in *Germany's Colonial Pasts: An Anthology in Memory of Susanne Zantop*, ed. Marcia Klotz, Lora Wildenthal, and Eric Ames (Lincoln: University of Nebraska Press, 2005) 148–164.

like Hegel's that presumes to transcend and "complete" the wisdom of scripture. He writes,

> ... in the face of the obvious ramblings, undertaken in the name of his sublime schemes, of someone who is probably the greatest thinker of all time, we ask ourselves whether language does not hold another secret to the one brought to it by the Greek tradition, and another source of meaning; whether the apparent and so-called "non-thought" representations of the Bible do not hold more possibilities than the philosophy that 'rationalizes' them, but cannot let them go free; whether the meaning does not stem from the scriptures that renew it; whether absolute thought is capable of encompassing Moses and the prophets—that is to say, whether we should not leave the System, even if we do so by moving backwards, through the very door by which Hegel thinks we enter it.[27]

In retrospect, we can see that, *Fear and Trembling* dramatically anticipates Levinas's advice. It seeks to "move backward" through Genesis 22, the door to the Jewish and Christian traditions. In doing so, it refuses to sunder one tradition from the other, and, by focusing on the individual—Abraham—it refuses to allow group identities of any sort to replace the constant need for individual choice and responsibility. The past two centuries have seen cultural anti-Semitism begin with the intellectual musings of a prestigious philosopher and end in genocide. In the wake of these events, *Fear and Trembling* merits a modest word of Jewish appreciation.

---

[27] *Difficult Freedom: Essays on Judaism*, 238.

8

Kierkegaard's Great Critique: *Either/Or* as a Kantian Transcendental Deduction

At first sight, it is hard to imagine two more different works than Kant's *Critique of Pure Reason* and Kierkegaard's *Either/Or*. Apart from their length and prolixity, these writings seem to have almost nothing in common. Kant's *Critique* is a painstaking epistemological study marked by precise logical analysis and aiming philosophically at architectonic completeness. *Either/Or* is a collage of disparate materials, lyrical aphorisms, literary reviews, personal memoirs by fictional characters, and essays with opposing views on esthetic and ethical subjects. If *Either/Or* contains a consistent subject matter throughout, it is erotic or romantic love—a theme not only totally absent from Kant's *Critique* but alien to Kant's life and philosophical interests.

Nevertheless, as Kierkegaard so often reminds us, one must be wary of appearances. Beneath the outer fabric of their concerns these two great works display a remarkably similar methodology. Working in the distinctive medium of romantic love, *Either/Or* employs the method of "transcendental deduction" that Kant had pioneered in the first *Critique* and thereafter had made the hallmark of his philosophical approach. Using this method, *Either/Or* undertakes a critique of *empiricism,* not just as an epistemological position, but also as a mode of life. Since it incorporates some of Kierkegaard's most basic concerns, *Either/Or* is also not just the first work in a series, but also, like Kant's *Critique,* a "propaedeutic" for an entire authorship. Taken together, these considerations warrant our regarding *Either/Or* as Kierkegaard's own great *Critique.*

To grasp these methodological similarities, it is important first to understand what Kant means by a transcendental deduction. Kant introduces three such deductions, one in each of his *Critiques,* but it is the deduction of the categories in *The Critique of Pure Reason* that represents this method in its strictest form. In a series of arguments in the portion of the *Critique* known as "The Transcendental Analytic," Kant shows that we cannot have the kind of empirical knowledge that we have unless we logically presuppose some knowledge that is not derived from experience. Specifically, Kant argues that our ability to effect a "synthesis of the manifold" of sense perception depends on a series of operations governed by a priori concepts that form the "categories" of pure reason. They include such concepts as unity, plurality, possibility, necessity, and causality (KRV 113).

In essence, a transcendental deduction is an effort to defend reason's claim to a priori knowledge against the skeptic's or empiricist's denial that such knowledge exists. A transcendental deduction begins with a cognitive experience shared by both sides to the dispute.[1] In the first *Critique,* this is the fact of our sensory knowledge of the world. No transcendental deduction can get under way without this reference to uncontested shared knowledge. The deduction then proceeds by showing that the mental operations underlying this cognition cannot be carried out unless another, and more fundamental, operation comes into play. In the case of the first *Critique,* this involves the employment of elements of cognition that shape experience but do not derive from it. Hence a transcendental deduction has as its distinctive feature a "not without" argument beginning with undisputed aspects of our cognition.

[1] Kant calls this direct, immediate awareness "a reflection."

One of the striking features of the transcendental deduction in the first *Critique* is the legal language that Kant uses to introduce it. In a well-known opening passage to the transcendental deduction, Kant mentions jurists' distinction between questions of right *(quid juris)* and questions of fact *(quid facti)*. Observing that only the former are given the title of a *deduction* (KRV 120), he then makes clear that his inquiry is a deduction in this sense because it will pursue the question of the *right* to knowledge.

Dieter Henrich situates Kant's understanding of a deduction in a legal tradition stretching back to the fourteenth century.[2] This tradition of "deduction writings" (*Deduktionsschriften*) originated as a way of settling land and succession disputes between the numerous petty rulers of the Holy Roman Empire. In most cases, a deduction sought to trace claims back to their origin. Kant was familiar with this tradition. Henrich believes that the entire first *Critique* and the way in which Kant presents his theory there, was "thoroughly affected by the decision to adopt juridical procedures as a methodological paradigm." He continues:

> The *Critique* is not just permeated by juridical metaphors and terminology. Its major doctrines are related to one another by means of the theory of legal disputes presented by Putter and Achenwall. A legal dispute originates when a party's claim has been challenged by an opponent, so that a court case must be opened. This happened in philosophy when the skeptic challenged the claim of reason to be in possession of a priori knowledge of objects. The dispute makes indispensable an investigation into the origins of such knowledge. To the extent to which a deduction can be produced, the claim of reason becomes definitely justified and

[2] Dieter Henrich, "Kant's Notion of a Deduction and the Methodological Background of the First *Critique*," in *Kant's Transcendental Deductions*, ed. Eckart Forster (Stanford, California: Stanford University Press, 1989) 32.

the challenge of the skeptic is rejected. This is the aim of the Transcendental Analytic.[3]

Understanding the legal/epistemological structure of Kant's employment of a transcendental deduction is a fitting introduction to Kierkegaard's effort in *Either/Or.* Some of the parallels between these works should immediately be obvious. As is true of Kant's transcendental deductions, Kierkegaard's argument takes the form of dispute between two parties that must be resolved. One, the young esthete A of *Either/Or,* Part 1, puts forth a position that denies the existence of any structures and possibilities that transcend the immediate moment of feeling and mood. The other, B, or Judge William of *Either/Or,* Part 2, defends the possibility and worth of a life based on the transtemporal category of duty. That the defender of the nonempiricist position is cast as a judge is noteworthy, since it underscores the adversarial, legal-deduction structure of the debate. Near the end of *Either/Or,* 2, the Judge makes the adversarial character of the book even clearer when he goes out of his way to remind us that he is not so much a judge as a "litigant" or "witness" in this dispute (EO2 322).

A deeper parallel appears when we look closely at the nature of the Judge's argument. A transcendental deduction, we have noted, proceeds from a cognitive experience shared by both sides to a dispute. For Kant in the first *Critique* this cognition, shared by empiricists and rationalists alike, is the fact of sensory intuition or *Anschauung*. Because we have an immediate, coherent experience of sensory objects, we must presuppose the a priori givenness in our cognition, first, of the realities of time and space, which are constitutive of any experience, and, second, of the conceptual categories through which the manifold of experience is unified. *Either/Or* is not primarily concerned with cognition. Its focus is on the esthetic, emotional, and ethical realities of life. To exhibit a

[3] Ibid., 38.

parallel to Kant's deduction, therefore, it must be organized around an element of esthetic-romantic-ethical existence that is as primordial as Kant's *Anschauung* and whose reality and nature are uncontested by both parties to the dispute. Such an element exists in *Either/Or*. It is *first love.*

First love (*første Kjerlighed*) is arguably the central theme of *Either/Or*. It makes its appearance as early as the *Diapsalmata,* one of whose aphorisms describes the first period of falling in love as "the most beautiful time ... when, from every encounter, every glance, one fetches home something new to rejoice over" (EO1 24). It reappears in the memoirs of Marie Beaumarchais, where it proves to be the source of her despair: "If love cannot last," she asks, "what, then, can last?" (EO1 187). It is the title of the play by Scribe reviewed by the young esthete, and its meaning is the play's underlying riddle. It is cynically dismissed in the "Rotation of Crops," and it is a propelling force in the "Diary of the Seducer," entering as the experience that initially attracts the seducer to his young victim and as the dynamic he exploits to work her eternally into his thrall.

As an experience, true first love is distinguished, above all, by its relation to time. As the Judge puts it at the beginning of his letter on "The "Esthetic Validity of Marriage," such love "bears a stamp of eternity." From the first moment that they see one another, the lovers are convinced that they have always been destined for one another and are meant to remain together forever. The seducer expresses this sensibility when he asks, "What does erotic love love? Infinity.—What does erotic love fear? Boundaries." The Judge agrees when he observes that all love, "whether it is superstitious, romantic, chivalrous love or the deeper moral, religious love filled with a vigorous and vital conviction, has precisely the qualification of eternity in it" (EO2 32). Elsewhere, he describes this sense of infinitude as "the apriority, that the first love has" (EO2 60).

At one level, it might seem that the esthete denies the reality of this transtemporal romantic experience. Certainly, much that he says mocks first love. His review of the Scribe drama, for example, ridicules Emmeline's equivocations over who is her first love, and the esthete presents her experience as characteristic of the confusions in this area. All talk about first love, he remarks, evidences a "sophistical thesis" because, when it is convenient, first love is presented *quantitatively,* as the first in a series. But at other times it is used *qualitatively* so that any intense infatuation, however late in a series, is offered up as a "first love." These confusions, the esthete observes, are laughingly displayed in the case of the widower and widow, each with five children, who "combine forces" and "nevertheless assure one another on the wedding day that this love is their first love" (EO1 254).

But this derision of first love is not the esthete's truest belief. No less than the young lovers in the romances he chronicles, the esthete admits to the reality of the experience of first love. He never denies that first love presents itself as unique and timeless, and he even exalts and pursues this experience for its own sake. The young seducer confesses that, although he "had not expected to be able to taste the first fruits of falling in love," he is now dazed by the experience and has "gone under in love-rapture" (EO1 324). The esthete's problem, here and everywhere, is that his basic understanding of human existence does not permit him to make sense of this experience in any way short of deriding it. If life is essentially about feeling, mood, and experience, as the esthete believes, there is no place for a feeling or experience that insists it will endure forever. Since all feelings pass, since change omnivorously devours all that exists in time, and since feelings and moods are fundamentally beyond our control, even this feeling of eternity must prove evanescent. Thus, it must be regarded as essentially spurious, and one should make every effort to avoid being ensnared and deceived by its allure. This can mean the avoidance of emotional involvement with others and a life lived on the purest plane of sensuous immediacy (Don Juan), or it can mean the deliberate, calculating seduction of another in a cynical and despairing effort to become master, rather than victim, of first love's power.

Judge William has no doubt that he and the esthete agree on the nature of the experience of true first love. Early in his essay on "The Esthetic Validity of Marriage," he offers a precise description of first love. It is, he says, "an absolute awakening, an absolute intuiting" [Danish *Skuen,* the cognate of Kant's *Anschauung*] that has as its object "a specific living being" (EO2 42). First love, he adds, "has an element of the sensuous, an element of beauty, but nevertheless it is not simply sensuous.... This is the necessity in the first love. Like everything eternal, it has implicit the duplexity of positing itself backward into all eternity and forward into all eternity" (EO2 42–43).

On the basis of this understanding, which he knows his young friend shares, the Judge is able repeatedly to convict the esthete of self-contradiction:

> [Y]ou are in pursuit of "the first"; you want only that, without suspecting that it is a self-contradiction to want the first to recur continually, and that consequently either you must not have reached the first at all or you actually have had the first and what you see, what you enjoy is continually only a reflection of the first.... (EO2 38)

> I beg you to recall the little contradiction we encountered—the first love possesses the whole content; to that extent it seems most sagacious to snatch it and then go on to another first love. But when one empties the first love of its content in this way, it vanishes, and one does not obtain the second *love* either.... (EO2 41–42)

> To be consistent, you must therefore hate all love that wants to be eternal love. You must therefore stop with the first love as a moment. But in order for this to have its true meaning, it must have an intrinsic naive eternity. Once you have learned that it was an illusion, it is all over for you, except insofar as you work to enter into the same illusion once again, which is a self-contradiction. (EO2 126–27)

The Judge knows that A cannot easily relinquish his commitment to the full significance of first love. As an esthetic experience in which the individual feels both exhilaratingly free and blissfully determined, both immersed in the particular and yet part of the universal (EO2 42, 90), first love is the consummation of the esthetic mode of life.[4] Were the esthete to renounce it, as his logic forces him to do, he would have to abandon all hope of fulfillment and give himself over to despair. To affirm first love in all its significance, however, to avoid what the Judge describes as the contradiction in the esthete's "whole being," he must move

---

[4] This idea of the bliss of romantic love as exhibiting a union of freedom and necessity suggests some of Kant's esthetic conceptions. Albert Hofstadter maintains that according to Kant's view of esthetic beauty, as developed in the *Critique of Judgment,* in the esthetic encounter with natural beauty and humanly created works of arts, "we experience an object which is felt and judged as freely necessary and necessarily free."—"Kant's Aesthetic Revolution," *Journal of Religious Ethics* 3, no. 2 (Fall 1975): 180.

It is difficult to determine to what extent Kierkegaard sought to incorporate aspects of Kant's esthetic theory into *Either/Or.* He was certainly familiar with Kant's third *Critique,* referring to it several times in his papers (e.g., JP4–4830 and 3853). The Judge's description of "the beautiful" as "that which has its teleology in itself" (EO2 272 and passim), sounds remarkably like Kant's conception of judgments of beauty as involving "disinterested satisfaction," a concept Kierkegaard refers to independently in JC 170. If Kierkegaard, in *Either/Or*'s final discussion of "The Balance between the Esthetic and the Ethical," sought to introduce aspects of Kant's third *Critique,* with its corresponding transcendental deduction of the idea of transcendental purposiveness in art and nature, this would strengthen my contention that *Either/Or* is meant to be a Kantian propaedeutic to Kierkegaard's entire authorship.

beyond esthetic categories. Having revealed the self-contradiction of an empirical-esthetic perspective, the Judge now proceeds to the positive side of his transcendental deduction: validating the presence of transtemporal and a priori elements in this human experience. In view of the conviction of eternity that marks first love, the Judge's task is clear. He must show that first love can endure. He reveals his own task when, speaking to the esthete, he remarks: "However you twist and turn at this point, you must admit that the task is to preserve love in time. If this is impossible, then love is an impossibility" (EO2 141).

The challenge before the Judge is considerable. In Kant's epistemological deduction in the first *Critique* the very immediacy of sensory intuition *(Anschauung)* provided certainty of the a priori elements of cognition, which had only to be philosophically clarified and developed. But substantiating the validity of first love requires going beyond the immediacy of its experience, since, if first love's inner meaning cannot logically be substantiated, it can always be dismissed as an illusion.[5] What the Judge must show, therefore, is that it is somehow possible to fulfill first love's inherent promise of eternity and to actualize and carry this promise out in time.

---

[5] In fact, Kant signals this difference between an epistemological and moral deduction in his second *Critique*. He notes there that a deduction relating to pure *practical* reason is much more difficult than one dealing with the principles of pure theoretical understanding. The latter concerns knowledge of the properties of objects "which can be given to reason from some other sources" (i.e., sensible intuition), whereas the former concerns knowledge "in so far as itself can become the ground of the existence of objects" (KPV 47–48).

Since they result from rational willing, in other words, the objects of pure practical reason are not given in experience but are brought into being by willing. Any deduction referring to them, therefore, proceeds from the structure and content of that willing and not from a given datum of sensory experience. In terms of first love, this means that any deduction makes reference to an object that is only present as an object of willing but whose real existence can always be negated in thought.

The key to the Judge's response is his sustained defense of marital love as the proper fulfillment of first love's promise. Within this defense, his argument proceeds along two lines. On the one hand, he tries to show that the "ought" of marital duty raises no bar to the continued presence of erotic and romantic love within marriage, that it is possible for married lovers continually to renew their experience of sensuous first love. On the other hand, he insists that first love is not only compatible with the "ought" of duty, but that it invites and requires it. This, in turn, involves showing that this "ought" is present within first love and that first love cannot exist without it. As we will see, Kantian insights permeate both aspects of the Judge's defense.

With regard to one line of argument, the continuance of the esthetic and emotional aspects of first love within marriage, the Judge's position is made difficult by the esthete's conviction that love and duty are opposed. Describing marital love as "absolutely unmusical" (EO1 83), the esthete places it at the furthest remove from sensuous immediacy. Love, as a spontaneous feeling, cannot be commanded, and when command appears, it must be either because the feeling has vanished or because it threatens to do so. In this spirit, the esthete counsels others to "never become involved in *marriage*," noting sarcastically that while married people pledge love for each other throughout eternity, they would be better advised just to say, "Until Easter, until next May Day." Despite these solemn pledges, he observes, "one of them detects after a short time that something is wrong, and then the other complains and screams: Faithlessness! Faithlessness!" (EO1 296) In a similar vein, the Judge renders an account, of the esthete's views when he remarks, "You say, 'within itself, marital love is hiding something completely different; it seems so gentle and beautiful and tender, but as soon as the door is shut on the married couple and before one can say Jack Robinson, out comes Master Erik[6]; then the tune is changed to duty'" (EO2 145).

Interestingly, the esthete's position here, though deeply rooted in popular thinking, can also be interpreted as deriving from a distorted but common reading of Kant's moral philosophy. Kant is well known for his insistence that even the most benevolent social feelings cannot serve as the determining ground *(Bestimmungsgrund)* of moral obligation. This is because feelings can never be commanded (and hence can never be a ground of moral imputation, praise or blame). This leads Kant to the conclusion that duty can be based only on willing, as this is guided by the (universal) faculty of reason. Since Kant, in his *Foundations of the Metaphysics of Morals,* had chosen to illustrate this point by holding up the case of the individual with an "unsympathetic nature" who nevertheless does his duty (GR 37, 56n. [398, 430]), it is common to believe that Kant deliberately opposed duty and emotion. But of course, in developing this pure case of action done undeniably "from" duty, Kant had no intention of barring from the moral life either social emotions or other forms of morally compatible sympathetic feelings. Although he always insisted on the strict priority of duty in shaping our conduct, he never claimed that this precluded one from simultaneously seeking to develop or pursue emotionally satisfying goals.

In some of his writings, Kant goes so far as to suggest that it is our moral *duty* to develop and nourish sympathetic and benevolent feelings towards others. For example, in a section of the "Doctrine of Virtue" of the *Metaphysic of Morals* dealing with "duties of love to other men," Kant states that such duties cannot, in the first instance, be taken to mean sympathetic emotions or feelings (what Kant calls "esthetic love") directed towards others.

[6] A decorated birch branch with which children awaken their parents on Shrove Monday. Here, presumably, the birch of discipline.

Instead, what is required is "practical love," in the sense of an active commitment to promoting their welfare. These ideas, of course, are a familiar expression of Kant's insistence that feelings cannot be commanded. However, Kant then adds the following observation:

> But while it is not in itself a duty to share the suffering (as well the joys) of others, it is a duty to sympathize actively in their fate; and to this end it is therefore an indirect duty to cultivate the compassionate natural (aesthetic) feelings in us, and to make use of them as so many means to sympathy based on moral principles and the feeling appropriate to them. It is therefore a duty not to avoid places where the poor who lack the most basic necessities are to be found but rather to seek them out, and not to shun sick-rooms or debtors' prisons and so forth in order to avoid sharing painful feelings one may not be able to resist. For this is still one of the impulses that nature has implanted in us to do what the representation of duty alone would not accomplish. (MS 250f.)

In a classic article, George Schrader has argued that the understanding of reason and inclination, duty and emotion, offered by Judge William in *Either/Or* represents a significant advance over Kant's perspective.[7] Kant, says Schrader, "virtually ignored" the role of consciousness and reason in the development of feelings and desires. For Kant, "Reason can impose an order on man's empirically given nature even though it cannot transform it from within." In contrast, "Kierkegaard views the sensuous erotic as developed through the agency of reason and imagination."[8] It should be clear, however, that Schrader's view is not based on a full reading of Kant. Not only is there nothing in Kant's view that requires such a strict separation of reason and feeling, but Kant

[7] George Schrader, "Kant and Kierkegaard on Duty and Inclination," *Journal of Philosophy* 65 (1968): 688–701.

[8] Ibid., 693–94.

himself also occasionally tried to correct this common misinterpretation of his position.[9]

Against this background, it is possible to interpret the Judge's campaign "to salvage the esthetic prestige of marriage" (EO2 6) as an effort, within a Kantian standpoint, to provide an alternative to this common misreading of Kant. For example, in affirming that "the esthetic, the ethical, and the religious" are "three great allies," and that "duty is always consonant with love" (EO2 147, 149), the Judge voices a position fully congruent with Kant's philosophy. He also works out Kant's broad injunction actively to develop our morally supportive emotions when he insists that marriage ideally is to have its point of departure in first love, and that the married pair should constantly strive to express, rejuvenate, and transfigure the esthetic-erotic-emotional side of their relationship (EO2 10, 31). In one respect, we can say that the Judge uses these deeper Kantian insights to advance his argument. In another respect, it is almost as though he uses this appreciation of marital life as a way of correcting an ingrained misreading of Kant.

The Kantian nature of the Judge's position becomes even clearer in the second of his lines of argument, his claim that first love is not only compatible with the "ought" of duty, but that it invites and requires duty. This is because duty is present in first love and because it is "not without" duty's presence that first love exists. Stated as such, we see here a *second* transcendental deduction in *Either/Or* (a deduction within his larger deduction) and one that exactly parallels the transcendental deduction that Kant offers in *The Critique of Practical Reason.* To understand Kierkegaard's moves here, it pays to look more closely at Kant's argument in the second *Critique.*

Kant's second major deduction is rather surprising. Since the first part of the second *Critique,* its "Analytic of Pure Practical

---

[9] For a fuller discussion of this, see my "Kant on Christian Love," (reprinted here as chapter two).

Reason," is devoted to the development of the idea of the moral law, we would expect that any deduction within this work would seek to deduce the reality of this law. This expectation is enhanced by the fact that Kant regards the law of duty, the categorical imperative, as something known a priori. Hence, we would think that just as the first *Critique* was devoted to providing a "deduction" of the a priori elements of cognition, so the second *Critique* would provide a "deduction" of the categorical imperative.

But this is not what Kant does. Instead, he announces that, "the moral law needs no deduction." This is because it is "an apodictically certain fact, as it were, of pure reason, a fact of which we are a priori conscious, even if it be granted that no example could be found in which it has been followed exactly" (KPV 48). Rather than being deduced, Kant continues, the moral law in its oughtness itself serves as a basis for a further deduction, that of moral freedom:

> Instead of this vainly sought deduction of the moral principle, however, something entirely different and unexpected appears: the moral principle itself serves as a principle of the deduction of an inscrutable faculty which no experience can prove but which speculative reason had to assume as at least possible (in order not to contradict itself....). This is the faculty of freedom, which the moral law, itself needing no justifying grounds, shows to be not only possible but actual in beings which acknowledge the law as binding upon them. (KPV 48–49)

The essence of Kant's position here, we can see, is a working out of the simple dictum that "ought implies can." It is also another example of the "not without" argumentative strategy of a transcendental deduction. The imperative of the moral law, Kant asserts, is an undeniable fact of our moral self-consciousness. But since nothing can be commanded apart from the presumption of our ability to pursue it, to the extent that we feel ourselves morally obligated, we must presume that we possess the freedom to obey the law's dictates.

It may seem a long way from this deduction in the second *Critique* to the Judge's defense of marital love, but the parallels between the two are striking. The Judge's defense involves three key ideas: that first love undeniably contains an "ought" to its continuance; that "ought implies can"; and that this implicit "can" is a consummate reason for celebration. Speaking to the esthete, the Judge reminds him of the imperative that first love contains, "Love is obviously not supreme for you," says the Judge, "for otherwise you would be happy if there were a power capable of constraining you to remain in it" (EO2 145). Marriage, in turn, by embodying this imperative fulfills the lovers' deepest wish:

> [M]arital love, ... in the ethical and the religious already has duty within itself, and when duty manifests itself to them it is not a stranger, a shameless outsider, who nevertheless has such an authority that by virtue of the secrecy of love one does not dare to show him the door. No, he comes as an old intimate, as a friend, as a confidant whom the lovers both know in the deepest secrecy of their love.... To them it would not be sufficient for duty to say encouragingly, "It can be done, love can be preserved"; but because he says: "It shall be preserved," there is an implicit authority that corresponds to the inwardness of their wish. Love casts out fear, but if love nevertheless fears for itself a moment, for its own salvation, then duty is precisely the divine nourishment love needs, for duty says, "Fear not; you shall *[skal]* conquer"—says it not in the future sense; for then it is only a hope, but in the imperative mood, and therein rests a conviction that nothing can shake (EO2 145).

The heart of Kant's second deduction is the employment of the givenness of moral duty—what he calls "the fact of reason"—to develop the reality of moral freedom. What I know myself inexorably to be required to do, I realize that I can do. The moral law and freedom thus "reciprocally imply one another" (KPV 29);

the moral law is the "ratio cognoscendi" of freedom while freedom is the "ratio essendi" of the moral law (KPV 4n.1). That Kierkegaard was familiar with Kant's arguments in the second *Critique* and that they were on his mind as he wrote *Either/Or* is suggested by a remark placed in the Judge's mouth that directly echoes Kant's deduction of moral freedom. Speaking of the ethically inclined individual, the Judge says that "when the passion of freedom is aroused in him—and it is aroused in the choice just as it presupposes itself in the choice—he chooses himself and struggles for this possession as for his salvation" (EO2 216).

It should now be clear that Kierkegaard's defense of the esthetic validity of marriage stands directly in this tradition of argument. Here, the "ought" of true first love's undeniable impulse toward constancy and fidelity is shown to be the ground for believing that this constancy can be maintained. Because they feel they *ought to* do so, the lovers know that they *can* preserve their love. True first love thus exists "not without" marriage and marital duty. With this deduction, the esthete's problem is solved. One need not succumb to the despair that results from seeing life's peak emotional experience as a snare and deception. By changing categories, by moving from the sphere of feeling to that of will—by means of "a determination of the will" solemnized in the marital vows—one can "rescue" love from the sensuous and realize its "true eternity" (EO2 21). A happily married man, the Judge concludes, has solved life's "great riddle." He is able "to live in eternity and yet to hear the cabinet clock strike in such a way that its striking does not shorten but lengthens his eternity" (EO2 138).

There is much else in *Either/Or* testifying to Kant's imprint on Kierkegaard's thinking. I have touched on some of these matters elsewhere.[10] Here I want to stress how *Either/Or* appropriates and applies Kant's major argumentative strategy in the *Critique of Pure Reason* and the *Critique of Practical Reason*. In view of the centrality of these transmuted transcendental deductions to the development of Kierkegaard's argument in *Either/Or*, it is no exaggeration, I think, to describe *Either/Or* as a deliberate *re-creation* and *re-presentation* of Kant's epistemological and ethical position in his first two *Critiques*. Subsequent works in Kierkegaard's authorship similarly track the progression of Kant's writings. Thus, the concern with sin and problems of soteriology developed by Kant in the second half of the *Critique of Practical Reason*, in the *Religion within the Limits of Reason Alone*, and in *The Conflict of the Faculties* are also re-created and re-presented in Kierkegaard's *Concept of Anxiety, Philosophical Fragments*, and *Postscript*. This parallelism of development raises a question I cannot pursue here but which an appreciation of the Kantian structure of *Either/Or* moves to the fore: To what extent did Kierkegaard deliberately develop his authorship on lines laid down by Kant's critical philosophy?

Whatever Kierkegaard's debt to Kant, *Either/Or* also makes apparent Kierkegaard's brilliant originality. While Kant's method of transcendental deduction may have provided Kierkegaard with his basic argumentative strategy, it was Kierkegaard who chose to apply the strategy to the dramatically different medium of romantic love. It is said that early New England church builders, by taking English conventions and applying them to the medium of wood, created a new architecture. Similarly, Kierkegaard, by transposing Kant's austere epistemological and ethical insights to the vital medium of life choices and human relationships, created a new philosophy.

---

[10] See my *Kierkegaard and Kant: The Hidden Debt*, (Albany: State University of New York Press, 1992) chapters three and four.

9

Either/Or: Kierkegaard's Overture

Even among the diverse formats and approaches found in Kierkegaard's pseudonymous writings, *Either/Or* stands out. It presents a collage of different contents and styles, ranging from the pithy and cynical reflections of the "Diapsalmata" to Judge William's book-length philosophical treatises. This diversity of formats and approaches coheres with the pseudonymous editor Victor Eremita's claim that the published volume represents his arrangement of a loose collection of papers that he had discovered in the hidden compartment of a second-hand writing desk.

Despite Eremita's comments, of course, *Either/Or* is by no means a haphazard collection of texts. Both in style and content, the book reflects careful authorial design. The young esthete's worldview shines forth in both the form and content of the materials of the first part and culminates in the narrative of a love-exploiting, love-destroying relationship of "The Diary of the Seducer." This provides occasion in the second part for the Judge to use romance and marriage as the springboard for his own defense of the ethical, while the Judge's confidence that one can live ethically, is called into question at the close by the country pastor's sermon.

Kierkegaard himself not only viewed the book as a coherent statement, but also perceived it as an appropriate start to his authorship. In *The Point of View for my Work as an Author*, he repeatedly describes the nearly simultaneous appearance in 1843 of the pseudonymous *Either/Or* and the signed *Two Upbuilding*

Discourses as the foundations of the dialectical structure of his total authorship. Furthermore, although Kierkegaard in *The Point of View* explicitly denies that he had a fully accurate idea of the authorship from the beginning, he acknowledges that he reflected carefully on every step he took along the way (POV 76f.).

What I would like to suggest here is that *Either/Or* was written not only as the first work in Kierkegaard's authorship, but also in some ways as a deliberate prelude to that authorship. More precisely, it serves as a keynote or overture to the authorship. Using the Latin word for work, *Either/Or* is the overture to Kierkegaard's *opera*.

Remarkably, *Either/Or* itself reveals that Kierkegaard gave considerable thought to what constitutes the appropriate prelude to a creative body of work. In the essay about Mozart's *Don Giovanni* entitled "The Immediate Erotic Stages or The Musical-Erotic," the young esthete spends a good deal of time specifically developing the concept of an operatic overture and distinguishing a good overture from a bad one. "The overture," he tells us there, "generally provides a profound glimpse into the composer and his psychical relation to his music." If the author "does not have a profound rapport with the basic mood of the opera, then this will unmistakably betray itself in the overture." Such a poorly crafted overture is merely "an assemblage of the salient points" but "not the totality that contains … the most penetrating elucidation of the content of the music" (EO1 126).

"The overture," the esthete continues, "frequently is a dangerous temptation for minor composers" who are "easily prompted to plagiarize themselves" and "steal from their own pockets." A good overture, he adds, "should not have the same content as the opera," but neither should it "contain something absolutely different." Instead, "it should have the same content of the opera, but in another way." It should grip the listener "with the full power of what is central" (EO1 126). Illustrating his point, the esthete draws our attention to the "perfect masterpiece" that is the overture to Mozart's *Don Giovanni.* "This overture is no mingling together [*Mellemhverandre*] of themes." It is "concise, defined, strongly structured, and, above all, impregnated with the essence of the whole opera." Furthermore, this is not attained "by sucking the blood of the opera; on the contrary, it is rather a prophecy in relation to the opera" (EO1 127).

One must be cautious about applying this recipe for a good overture to Kierkegaard's achievement in *Either/Or*. The esthete himself tells us "it is appropriate that the overture is composed last so that the artist can be really saturated with the music." Obviously, *Either/Or* is not a deliberate summation of Kierkegaard's work because it predates it. Nevertheless, like the good overture the esthete describes, whose "intended effect is to evoke a mood," *Either/Or* does provide a lyrical anticipation of the pseudonymous writings that follow. It accomplishes this, moreover, without "sucking the blood" from those writings. Although some of the leading themes of Kierkegaard's subsequent works make their appearance, and even in some cases are partly developed in *Either/Or,* these treatments are only suggestive. Even the illustrations used are very different from those offered in the later writings. Like a good overture, *Either/Or* "sets the mood" for and prophecies what follows, but it avoids creating a situation where any part of the authorship plagiarizes another.

Much of what I am saying will be familiar to students of Kierkegaard who have long perceived important harbingers of the authorship in *Either/Or*. What I want to do, however, is tease out a few of the major themes that make their appearance here to illustrate just how well *Either/Or* manages to evoke the mood of the authorship without anticipatorily stealing anything from the later writings. I organize this brief overview in terms of some of the key themes emerging in later works that make their first appearance in discreet sections of *Either/Or*. Within the latter, I particularly want to focus on the description of the modern Antigone that appears in *Either/Or* I and Judge William's long letter/discourse on the "Esthetic Validity of Marriage" in *Either/Or* II.

The Modern Antigone

Immediately following his treatment of Mozart's *Don Giovanni,* the esthete offers a lecture on "The Tragic in Ancient Drama Reflected in the Tragic in Modern Drama." Embedded within this lecture is an imaginative "modern" re-invention of the *Antigone,* a reframing of the ancient drama in which "everything is the same and yet everything is different" (EO1 154). As in the Greek drama, Oedipus is hailed and admired by the Thebans, whom he has liberated, and he is happy in his marriage with Jocasta. But in a deviation from the original, Antigone alone knows about her father's past: that he has killed his father and married his mother. This secret is Antigone's burden. While the tragedy of the Greek heroine lies in a young woman's being destined to be "alive in a place of corpses, never at home with the living nor the dead," the modern Antigone must say this of her entire life, since she bears within her heart an awful secret that she can reveal to no one. As a result, her life "is essentially at an end" (EO1 156).

The modern Antigone cannot even break out of her solitude by commiserating with her father. "She loves her father with all her soul, and this love draws her out of herself into her father's guilt. As the fruit of such a love she feels alien to humankind. She feels her guilt the more she loves her father." Antigone would gladly confide in Oedipus. But not only can she not do this now that he is dead, "but while her father was living, she could not confide her sorrow to him, for she indeed did not know whether he knew it, and consequently there was the possibility of immersing him in a similar pain" (EO1 161). In the esthete's imagination, this tormenting solitude pursues Antigone throughout her most intimate relationships. He imagines her as falling in love, but in that case "her dowry is her pain." "Without this dowry, she cannot belong to any man." To conceal this past "would be impossible; to wish to have concealed it would be a breach of her love—but with it can she belong to him?" (EO1 163). Antigone "struggles with herself; she has been willing to sacrifice her life for her secret, but now her love is demanded as a sacrifice." To this is added one further collision: her sympathetic love for her beloved. "With every protestation of love, he increases her pain; with every sigh, he plunges the arrow of grief deeper into her heart.... He beseeches her in the name of the love she has for her father ... placing all his hope in this means, not knowing that he has actually worked against himself" (EO1 164).

Close readers of Kierkegaard's work will see that while this modern Antigone cannot wed, her story is nevertheless pregnant with many of the ideas developed later in the authorship. For one thing, she anticipates the several ill-starred lovers depicted in *Fear and Trembling* who silently bear a tragic secret. Like the young swain whose decision to wed could destroy a family, or like Sarah of the *Book of Tobit* all of whose previous grooms have perished in the bridal chamber (FT 85, 102–107), the modern Antigone inherits a legacy of familial misfortune that blocks the way to marriage. Like them, she is caught between the ethical mandate to reveal everything to the beloved and the compassionate instinct to spare the beloved suffering, an instinct that in some cases requires the necessary cruelty of breach, deception, and concealment.

The narratives of tragic love in *Fear and Trembling* are central to the book's argument. On one level, they are used to illustrate the pseudonymous author's point about the required communicability of the ethical (in contrast to the incommunicability of its teleological suspension). A lover esthetically defies the ethical by failing to communicate with the beloved. Abraham religiously transcends the ethical by abstaining from open discourse with others. As Abraham's impassioned defender Johannes de Silentio reminds us (and as his name suggests), silence and solitude may have a place in esthetics and are hallmarks of faith, but they are forbidden by the ethical.

But these stories about ill-starred lovers in *Fear and Trembling*'s third Problema also have a much deeper relationship to the meaning of the text. In a series of published articles I have repeatedly argued that *Fear and Trembling* is *not* about ethics.[1] By this, I mean to say that it is not primarily aimed at depicting and justifying a suspension of paternal (or other) moral duty in the name of obedience to God. Rather, it is a text, which, in a long tradition of Christian writing, uses Abraham as a "figure" for the God-Christ dynamic of salvation. On this account, the "teleological suspension of the ethical" points not to the human suspension of moral duty, but to God's suspension of humans' deserved punishment for sin through the atoning sacrifice of God's own beloved son. *Fear and Trembling,* in other words, belongs more to the literature of soteriology than it does to ethics.

When *Fear and Trembling* is read primarily as a text dealing with the theme of human sin and its forgiveness, many of its specific features assume a different meaning. For example, the relationship to Kierkegaard's own life, usually seen as an undercurrent in the text, undergoes significant change and a deepening in meaning. When the primary theme is taken as the tension between human moral duty (in the form of family obligations) and the relationship to God, the text points to Kierkegaard's breaking of his engagement to Regine in order to pursue his vocation as a religious author. But when sin is the

[1] Ronald M. Green, "Deciphering *Fear and Trembling*'s Secret Message"; "Enough Is Enough! *Fear and Trembling* is Not about Ethics," and "A Reply to Gene Outka"; "'Developing' *Fear and Trembling*" (reprinted here as chapter six).

issue, then the breach with his fiancée becomes only the consequence of a much deeper problem: the Kierkegaard family's own tradition of sin, begun by Kierkegaard's melancholy father, Michael Pedersen Kierkegaard. As the book's epigraph suggests, *Fear and Trembling* then becomes a secret message between father and son about their mutual involvement in sin and about the possibility—indeed, the hope—that God's gracious forgiveness can redeem even such a tragic lineage. On this reading, Regine Olsen becomes a secondary secret reader who learns not that she was dropped in the name of a life devoted to God, but that Kierkegaard could not dare to involve her in this dark family history of melancholy and sin. On this reading, Regine is not Isaac to Søren's Abraham. Rather, Søren is the sacrificed but possibly redeemed son while Regine is both a bystander and further victim of this family drama.[2]

This reading also makes much deeper sense of the repeated tales of obstructed love in the book. It is not just the fact that something intervenes to prevent the consummation of a union that is noteworthy. Rather what intervenes in almost all these cases—from the tale of Sarah and Tobias to that of Agnes and the merman—is a destructive prior history of death or sin. These tales of impossible love, then, are not just asides introduced to make some subordinate point about ethical silence or openness. They are essentially related to the Kierkegaardian family drama that animates the book's main preoccupation: sin and its forgiveness.

This returns us to *Either/Or*'s modern Antigone. What initially seemed to be an imaginative discourse on the nature of tragedy now reveals itself as a window into the world of concerns that animates Kierkegaard's authorship from start to finish. How

[2] For a treatment of the importance of Søren's relationship to his father, including its connections to Søren's thinking about hereditary sin, see Joachim Garff, *Kierkegaard: A Biography* (Princeton: Princeton University Press, 2000) 346f.

is sin to be comprehended? How is it occasioned and transmitted? What are its consequences for the inner life of the individual and for his relationship to others, especially to intimates? What is the role of silence and solitude in the life of a sinner and of the religious individual? And yet, while this brief treatment of Antigone hints at all these questions, it does not answer them. Kierkegaard's overture does not plagiarize the opera. The lyric note is sounded for those willing to listen, but the longer arias lie ahead.

The connection of these passages in *Either/Or* to Kierkegaard's larger authorship becomes even clearer when we note how prevalent within the treatment of Antigone are the themes of hereditary sin and anxiety. These themes, of course, are picked up and greatly expanded in *The Concept of Anxiety*, but they are also suggested here. For example, the esthete uses the *Antigone* story to identify a further similarity and a difference between the ancient and modern narratives. Both heroines are shaped by their family's history. The outer train of events, the inevitable workings of fate initiated by her father's deeds, crushes the classic heroine. "In her childlike piety the Greek Antigone participates in her father's guilt." For her, however, "the father's guilt and suffering are an external fact, an unshakable fact that her sorrow does not move. But for our Antigone it is different." Her love for her father actively implicates her in his sin and makes her a willing participant in it. Whether because she "did not have the courage to confide in him," or because "she is continually in conflict with her surrounding world," she does not merely inherit his sin, in the sense that she falls victim to his guilty deeds. Her life is a willed recovery and reappropriation of his guilt.

Ettore Rocca has significantly amplified the possible meaning of the esthete's discussion by drawing our attention to the sexual motifs implicit in the esthete's treatment of Antigone. "Antigone," Rocca observes, "is the bride of her dead father's recollection and, in her sorrow, she expresses her love for him. In this love there is almost the symmetrical guilt of Oedipus: Antigone is the bride of her father's sorrow, i.e. in inwardness she is the bride of her father. Therefore she becomes 'equally guilty' as Oedipus, guilty of the same crime: incest; a modern incest, of course, an incest of reflection, but still an incest. And the fruit of this love must be kept secret and hidden from the eyes of all, because it is the sign of the deepest possible guilt."[3]

The modern Antigone is also familiar with anxiety. "At an early age, before she had reached maturity, dark hints of this horrible secret had momentarily gripped her soul, until certainty hurled her with one blow into the arms of anxiety." Anxiety, the esthete explains, "is the vehicle by which the subject appropriates sorrow and assimilates it." It is "the motive power by which sorrow penetrates a person's heart." As an erotic glance craves its object, "so anxiety looks cravingly on sorrow." But unlike lust or love, "anxiety has an added factor that makes it cling even harder to its object, for it both loves and fears it" (EO1 154f.). Anxiety also contains a reflection on time. "I cannot be anxious about the present but only about the past or the future." The past, in the form of her father's unfortunate fate rests like an "impregnable sorrow" on her, and it is also the source of the forebodings about her own fate. The modern Antigone's Greek counterpart also sorrows, but her sorrow is about the present. It is a deeper sorrow, "but the pain is less" (EO1 155).

Readers of *The Concept of Anxiety* will see here anticipations of that work. Antigone's ambivalent attraction to/repulsion from her sorrow becomes in the latter work anxiety's "sympathetic antipathy and antipathetic sympathy" (CA 4). Anxiety's relationship to past and future takes form in *The Concept of Anxiety* as "freedom's

[3] Ettore Rocca, "The Secret: Communication Denied, Communication of Domination," in *Søren Kierkegaard and the Words: Essays on Hermeneutics and Communication*, ed. Poul Houe and Gordon D. Marino (Copenhagen: C.A. Reitzel, 2003) 122.

possibility." In *The Concept of Anxiety* we learn that, like the modern Antigone, we can be anxious about our past, because a past that is not once and for all repented always "stands in a relation of possibility to me." I can be anxious about a past "because I have not placed it in an essential relation to myself as past and have in some deceitful way or another prevented it from being past" (CA 91f.).

Compared with *The Concept of Anxiety, Either/Or*'s treatment of anxiety in connection with Antigone is only a promissory note, only a hint. In his discussion of Mozart's overture, the esthete particularly draws our attention of the sensuous-erotic motif that represents Don Giovanni. The esthete remarks that, "the beginning of it is admirably expressed. We hear it so faintly, so cryptically suggested. We hear it, but it is over so swiftly that it is as if we had heard something we had not heard. It requires an alert ear, an erotic ear, to notice the first time the hint is given in the overture of the light play of this desire that is so richly expressed later in all its lavish profusion" (EO1 128). Nils Holger Petersen believes that he is able to identify the precise point in the overture to which the esthete refers. He believes that this is the anxious set of violin motifs occurring from bar eleven to bar fourteen of the slow D minor introduction to the overture, motifs that reappear much later in Don Giovanni's surprised response in the second act to the arrival of the statue of the Commendatore. For Petersen the association of this motif with the Commendatore confirms the esthete's view that Don Giovanni's sensuality betrays anxiety.[4] The same could be said of the motifs sounded only briefly in the treatment of the modern Antigone. It requires an alert ear, an ethico-religious ear to perceive them. Bringing

[4] Nils Holger Petersen "Søren Kierkegaard's Aestheticist and Mozart's *Don Giovanni*," in *Interart Poetics: Essays on the Interrelations of the Arts and Media*, ed. Ulla-Britta Lagerroth, Hans Lund, and Erik Hedling (Amsterdam: Rodopi, 1997) 167–176.

together in a single illustration the key Kierkegaardian themes of hereditary sin, sexual transgression, spiritual isolation, silence, and the relationship between time and eternity, this brief discussion in *Either/Or* sets the tone for the works to come.

By displaying the essential relationship between all these ideas, the treatment of Antigone also provides a key to understanding them as they are later developed. Gordon Marino has observed that *The Concept of Anxiety* is a "maddeningly difficult book," in which, he says "there are many passages ... the meaning of which completely escapes me."[5] Many of us have shared Marino's frustration with this difficult work. By listening carefully to Kierkegaard's overture at this fleeting moment, however, we can gather additional insight for the score that follows.

The "Esthetic Validity of Marriage"

In his review of *Don Giovanni*, the esthete comments on the movement or progression of the opera's overture. It begins, he says, "with a few, deep, even earnest notes." These signify the Commendatore. But the interest of the opera is Don Giovanni alone, "not Don Giovanni and the Commendatore." For this reason, "Mozart seems to have deliberately designed it in such a way that the deep voice that rings out in the beginning gradually becomes weaker and weaker." It "must hurry to keep pace with the demonic speed that evades it" and that "gradually creates the transition to the opera itself" (EO1 127). In this respect, the overture is like a sunrise. "So it is in nature that one sometimes sees the horizon dark and clouded," hiding everything in the obscurity of night. "Then in the most distant heavens, far off on the horizon, one sees a flash." This slowly gathers strength until it

[5] Gordon Marino, "Anxiety in *The Concept of Anxiety*," in *The Cambridge Companion to Kierkegaard*, ed. Alasdair Hannay and Gordon Marino (Cambridge: Cambridge University Press, 1998) 308.

begins to illuminate the entire heaven with its flame and "it seems as if the darkness itself has lost its composure and is beginning to move" (EO1 129). Don Giovanni represents "the full force of the sensuous." After Mozart has had him come into existence this way, "his life now develops for us in the dancing strains of the violin, in which he lightly, fleetingly speeds on over the abyss ... jubilating during his brief span" (EO1 129f.).

Implicit in this description of the overture is the idea that a good overture should, in its development and progression, evidence the dominant theme of the opera, and, once again, Kierkegaard seems to have heeded the esthete's advice. The directionality of *Either/Or* also betrays the dominant motif of the authorship as a whole. But here the progression is in exactly the opposite direction of *Don Giovanni.* Since Mozart's opera epitomizes the musical-erotic, it moves from moral gravity—earnestness—to utter sensuousness. Don Giovanni's voice comes to eclipse that of the Commendatore. But in *Either/Or,* it is the light, cynical, bantering voice of the esthete that is progressively overcome and eclipsed by earnestness.

This progression is already evident in the papers of the esthete. Beginning on a tone of haughty cynicism in the Diapsalmata, they conclude on the last pages of the "Seducer's Diary" with a note of revulsion. The esthetic approach to life has lost its charm. The love object has utterly "lost her fragrance" (EO1 445).

The way is thus prepared for the devastating critique of the purely sensual-erotic life in the Judge's treatment of "The Esthetic Validity of Marriage." Elsewhere, I have argued that this treatise seems be modeled on a Kantian transcendental deduction.[6] Like such a deduction, it takes a given aspect of empirical experience and shows that that aspect cannot be given in experience without the presupposition of some prior, non-empirical, conceptual

[6] See chapter eight.

reality. In Kant's case, this "not without" argument is used repeatedly in his writings to vindicate the a priori nature of space and time, the necessity of the categories of cognition, the moral law, and human freedom. In the case of Judge William, the argument proceeds not from cognitive/moral experience but emotional/moral experience. The Judge focuses on that which the esthete repeatedly admits is the pinnacle of sensuous-emotional life: the experience of first love.

In "The Esthetic Validity of Marriage," the Judge demonstrates that the fulfillment of the sense of eternity in sensual erotic love cannot occur without a movement from a life based on a passive response to satisfactions and moods to one based on active ethical commitment and resolve. The esthetic sphere of existence points to its successor, first love points to marriage, but the transition is made not by staying within esthetic presuppositions but only by a willed decision to replace enjoyment with imperturbable ethical resolve. One cannot reach the certainty of love's persistence by what Kierkegaard later describes as an "approximating transition,"—by some ever-greater refinement of mood or feeling. One can only do so by a qualitative shift in one's guiding premises, by a leap.

In this ethical resolve, the married person also conquers time. Not only can love be made to abide, but time itself is also recovered and permeated with the eternal. No longer the "dangerous enemy" of human finitude, time now lends meaning to human life. The infinite is experienced in time and our experience in time discloses the infinite. The married man, the Judge tells us, "has not killed time but has rescued it and preserved it in eternity." He "truly lives poetically" and "solves the great riddle, to live in eternity and yet hear the cabinet clock strike in such a way that its striking does not shorten but lengthens his eternity" (EO2 138).

Readers of Kierkegaard will see all the motifs introduced here as crucial to his authorship as whole. The Judge's deduction of love's need for marriage provides the authorship's first illustration of the leap as the "category of decision" (CUP1 99). While *Fear and Trembling* and the *Postscript* will take these concepts to new heights and apply them across the whole compass of existence spheres, the Judge's argument here provides a glimpse into the basic dynamics that Kierkegaard develops into a philosophy of existence. Indeed, at the very close of *Either/Or*, the country pastor, by calling into question the Judge's own tranquil confidence in his ethical integrity and fidelity, will suggest the religious leap that becomes the principal focus of *Fear and Trembling* and the *Postscript*. Judge William has already sounded the note and set the mood.

The theme of time and its relationship to eternity that will become a leitmotif of the authorship also makes its appearance, but in an inverted form. Here the problem is how eternity can be experienced in time, how eternity can recover time, how love can survive the ravages of aging and death. In the *Fragments* and the *Postscript*, however, the problem is how an event in time, specifically the death and resurrection of Jesus Christ, can be thought essential to an eternal destiny. In the words of the *Fragments*' epigraph, the questions are: "Can a historical point of departure be given for an eternal consciousness; how can such a point of departure be of more than historical interest; can an eternal happiness be built on historical knowledge?" (PF 1). Despite this different angle of approach, however, the underlying problem is the same. How can human beings, as creatures living at the intersection of time and eternity, make sense of our lives? How can we validate both time and eternity without fleeing from the eternal into the temporal or from the temporal into the eternal?

On this point I will conclude. There is much more that could be said. Woven throughout the texts of *Either/Or* are many more pointers to the authorship to come. For example, the theme of despair, not fully developed until near the end of the pseudonymous authorship in *The Sickness unto Death,* makes a repeated appearance even in this early work. We find it in the tale of the "Unhappiest Man," who, in not being able to die, not being able to "slip down into a grave" (EO1 220), offers a premonition of the later work's assertion that "the torment of despair is precisely the inability to die" (SUD 18). And in the Judge's assertion that the person who wills despair "is truly beyond despair" (EO2 213), we find an anticipation of Anti-Climacus's paradoxical assurance that "it is the worst misfortune never to have had that sickness" and "a true godsend to get it" (SUD 26).

In his treatment of the overture of *Don Giovanni,* the esthete remarks "To anyone hearing the overture after he has become more familiar with the opera, it may seem as if he had penetrated the hidden workshop where the forces he has learned to identify in the opera move with a primitive power, where they wrestle with one another with all their might" (EO1 127). Much the same can be said of *Either/Or*. Although a youthful work in every sense, it boldly anticipates the sophisticated body of work to follow. Re-reading it from the perspective of the later work, we truly feel that we have "penetrated the hidden workshop" of Kierkegaard's creative endeavor.

10

Erotic Love in the Religious Existence-Sphere[*]

Works of Love has often been regarded as a religious discourse devoted to the theme of Christian love. Throughout, Kierkegaard draws a contrast between Christian love, which aims at the good of every fellow human being, and the particularized forms of merely human attachments: friendship, erotic love, or marriage. Because of this contrast, it is easy to conclude that Kierkegaard's vision of the religious sphere of existence holds no place for erotic love, with its passionate intensity and particular focus. However, there are repeated suggestions in *Works of Love* that Christian faith transforms but does not replace special love for another human being, just as marriage transforms but does not eliminate erotic attraction to another person. These suggestions are important, because they indicate that *Works of Love* is not just a treatment of Christian love but also, in some ways, the culmination of a comprehensive philosophy of love that is developed by Kierkegaard in the course of his writings.

Kierkegaard's attention to the theme of particularized or erotically charged love between persons is one of the more neglected aspects of his work. Insightful discussions of *Works of Love*, such as those by Gene Outka or Gilbert Meilaender, do not usually link this work to Kierkegaard's treatments of erotic love.[1]

---

[*] This chapter was co-written with Theresa M. Ellis.

[1] Gene Outka, *Agape: An Ethical Analysis* (New Haven, Connecticut: Yale University Press, 1972) 13–24; Gilbert Meilaender, *Friendship: A Study in Theological Ethics* (Notre Dame, Indiana: University of Notre Dame Press, 1981) 7, 27–28, 35, 42–47, 53, 62–64, 84.

In turn, scholarly examinations of "love" in Kierkegaard's writings report the views of his pseudonyms on erotic and married love, but few seek to present these as a coherent position or to integrate them with Kierkegaard's discussion of Christian love in *Works of Love*. For example, Irving Singer, who includes a discussion of Kierkegaard within his broad philosophical treatment of love, focuses on *Works of Love* in terms of the contrast Kierkegaard draws there between particular attachments, especially erotic or romantic love (*Elskov*) and Christian love (*Kjerlighed*). According to Singer, by the time Kierkegaard reaches *Works of Love*, "all idealizations of romantic love have been discounted. Though hallowed by religious love, marriage no longer functions as a vehicle toward the ideal. The ethical universal [of marriage] has been swallowed up by the love of God. All other values are subsumed under that alone."[2] In reaching this conclusion, Singer mentions Kierkegaard's extensive treatments of erotic love and marital love only in passing. Recently, some writers have begun to link *Works of Love* to Kierkegaard's other treatments of love and male-female relationships generally.[3] But these emergent discussions, often in the context of other interests, have not fully highlighted how important themes of love are for Kierkegaard's authorship as a whole.

---

[2] Irving Singer, *The Nature of Love*, 3 vols. (Chicago: University of Chicago Press, 1984) 3: 47.

[3] See, for example, Sylvia Walsh, "Forming the Heart: The Role of Love in Kierkegaard's Thought," in *The Grammar of the Heart*, ed. Richard H. Bell (San Francisco: Harper & Row, 1988); also, her *Living Poetically: Kierkegaard's Existential Esthetics* (University Park: The Pennsylvania State University Press, 1994) 99–109, 181–88, 262–66; and her "On 'Feminine' and 'Masculine' Forms of Despair" in *Feminist Interpretations of Søren Kierkegaard* 203–215; M. Jamie Ferreira, "Equality, Impartiality, Moral Blindness in Kierkegaard's *Works of Love*," *Journal of Religious Ethics* 25, no. 1 (1997): 65–85; and Amy Laura Hall, "Poets, Cynics and Thieves / Vicious Love and Divine Protection in Kierkegaard's *Works of Love* and *Repetition*," *Modern Theology* 16, no. 2 (April 2000): 215–36.

This relative neglect is unfortunate. Despite its obvious importance in human life, erotic love between human beings has been little studied by philosophers or theologians. There are a handful of classical treatments, of which Plato's *Symposium* is perhaps best known. In the modern era, a small number of philosophers—among them Singer, Robert Solomon, and Laurence Thomas[4]—have sought to analyze the experience of being in love with another person. Within this relatively meager philosophical tradition, therefore, Kierkegaard's contribution bulks large. He devotes entire sections of *Either/Or, Fear and Trembling, Repetition, Stages on Life's Way,* and *Two Ages* to this topic, and he and his pseudonyms repeatedly discuss the similarities and differences between erotic love, married love, and Christian love throughout the pseudonymous and devotional writings.

Kierkegaard's treatments of love also provide a major tool for understanding his thinking as a whole. His writings repeatedly offer instances of human-to-human love as illustrations of larger esthetic, ethical, and religious truths. For example, in *Fear and Trembling,* Johannes Climacus uses the fate of a star-crossed lover who finds himself hopelessly in love with a princess to illustrate the psychological dynamics and decision making of a "Knight of Infinite Resignation" (FT 41–44). In *Either/Or,* Judge William's extensive reflections on his marriage become a means of illustrating the contrast between the esthetic and ethical spheres of existence. Kierkegaard's writings can be said to use human-to-human love (as Saint Paul does in Ephesians 5:31–32) to point readers toward higher, spiritual truths. Thus, Kierkegaard's philosophy of love is an important contribution to our thinking

---

[4] Singer, *The Nature of Love;* Robert C. Solomon, *About Love: Reinventing Romance for Our Times* (Lanham, Maryland: Rowman & Littlefield, 1994); Laurence Thomas, "Reasons for Loving," in *The Philosophy of (Erotic) Love,* ed. Robert C. Solomon (Lawrence, Kansas: University Press of Kansas, 1991) 467–476.

about this under-examined subject, and it also furnishes a useful point of entry into his authorship as a whole.

To develop fully Kierkegaard's philosophy of love is a separate task that still awaits doing. Our aim here is a much more limited: to provide a brief sketch of some of the main elements of this philosophy as a way of illuminating the conception of erotic and marital love developed in *Works of Love*.[5] Superficially regarded, *Works of Love* can be seen as presenting Christian love in dramatic contrast both to friendship and to particularized erotic love for another human being in any of its forms: whether married or unmarried. To fully understand Christian love, therefore, we must at least know what are the common features of its precursor forms. In a deeper sense, however, we will try to show that in *Works of Love* (and the writings on love that precede it) erotic love for Kierkegaard is less an anti-type to Christian love than a driving force that leads to it. Propelled from crisis to crisis by its own inherent tensions and contradictions, erotic love moves from tumultuous esthetic expression, through the challenges of married love, until it ultimately finds rest and peace in the religious sphere of existence. To understand the nature of Christian love and why it ultimately transforms all preliminary forms of love, therefore, we must examine what Kierkegaard (or his pseudonyms) has to say about the nature of love in its evolving expressions.

This discussion has three parts. First we propose to identify the features that mark erotic love in all its forms. In fact, these features are present in every form of love, including Christian-religious love, although one feature—exclusivity—may not initially appear to be so. Second, we wish to develop the way by which love serves as a driving force that leads individuals,

[5] Some of the material in this discussion is drawn from Theresa M. Ellis's Dartmouth College Honors Dissertation, "'A Kiss Which Was Something More than a Peck': Kierkegaard's Embrace of Romantic and Erotic Love in his Pseudonymous and Religious Writings." (Hanover, New Hampshire, 1997).

dialectically, from one sphere of existence to another. Finally, drawing on a series of suggestive comments by Kierkegaard in *Works of Love,* we hope to evidence the ways that erotic love reaches its culmination and fulfillment in the religious sphere of existence.

Features of Erotic Love

In the corpus of Kierkegaard's writings, we can identify a series of features that he or a pseudonym offers as common to erotic love in its esthetic expressions and its form as married love. Although Kierkegaard sometimes applies the term erotic love (*Elskov*) only to the more purely esthetic expressions of love, we will use the term "erotic love" for all the forms of person-to-person, physical-emotional attraction in which these features are found. Used this way, erotic love also refers to "romantic love" (*romantiske Kjerlighed*) and "marriage" (*Ægteskab*), terms that Kierkegaard employs to denote particular varieties of erotically charged male-female relations.[6]

Much could be said about each of the six features we bring to the fore: love's passionate nature; its exclusivity; its non-rationality; its tendency toward equality; its tendency to unite freedom and necessity as well as time and eternity. To some extent the persistent emphasis of these features in Kierkegaard's direct and pseudonymous writings reflects a kind of phenomenological analysis of the experience of love. In other respects, their presence in Kierkegaard's writings betrays the influence of generations of literary explorations of love, the proximate influence of the Romantic movement, and features of Kierkegaard's own complex biography, including his broken engagement to Regine Olsen. A

---

[6] Strictly speaking, nothing in Kierkegaard's conception of person-to-person erotic love necessitates its being heterosexual, but obviously Kierkegaard was very much a child of his time in failing to develop his philosophy of love in other than heterosexual ways.

full account of Kierkegaard's thinking about love would have to comprehend the many influences and assumptions that go into it. We put these matters aside in order to provide here a brief overview of the position we find in his writings

1. Love is a passion

Kierkegaard presents passion (*Lidenskab*) as a driving desire that consumes an individual and pushes her or him onwards towards a goal. It can take many forms. In *Philosophical Fragments,* for example, Johannes Climacus tells us that reasoned thought has a passion that involves never being satisfied with a superficial explanation and always seeking to determine the essence of things. This eventually takes the form of driving the understanding "to want to discover something that thought itself cannot think" (PF 37). Religion, too, is marked by passion. "Behold, faith is indeed the highest passion of subjectivity," Climacus tells us in the *Postscript* (CUP1 132), and later adds, in a memorable definition, "Faith is the contradiction between the infinite passion of inwardness and the objective uncertainty" (CUP1 204). Passion, finally, is an omnipresent feature of erotic love in all its forms. "To love without passion is an impossibility," Kierkegaard tells us in *Works of Love* (50). In *Either/Or* the esthete tells us that erotic love develops quickly, opening for the lovers "with a snap as the passion-flower does" (EO1 241). Elvira, that consummate example of erotic attachment, stands out for "the staggering intensity of her passion" (EO1 191). Despite its potential for abuse and his disclaimer that he has no intention to sanction "every unshaven passion" (JP3–3127), Kierkegaard seems largely to regard passion in positive terms. He considered a lack of passion as one of the great shortcomings of his era, what he termed its "tragedy of reason and reflection" (JP3–3129). Frater Taciturnus, in *Stages on Life's Way*, remarks that "The same thing that weakened faith in love—the lack of a sense of the infinite—will also weaken faith in the other passions" (SLW 410). In view of this, we can understand why erotic love should play such an important role in Kierkegaard's thinking and why instances of erotic love abound in his writings as metaphors for religious faith.

2. Love is exclusive

The whole passion of erotic love is focused on a single individual: "[E]rotic love intensifies in the direction that there is but one and only one beloved" (WL 49); "Passion always has this unconditional characteristic—that it excludes the third...." (WL 50). True love also is exclusive across time: it permits the love of only one person in a lifetime. To recoil from a failed love affair into a new relationship, is not to fall in love for a second time, says Kierkegaard, but to admit that the first love really was no love at all. In erotic love, Kierkegaard tells us, "there is but one and only one beloved in the whole world, and this one and only one time of erotic love, is everything; the second time is nothing. Ordinarily one says proverbially that just one time does not count. Here, in contrast, one time is unconditionally everything and the second time is unconditionally the ruin of everything" (WL 49). This is the reason why frustrated but genuine loves, in Kierkegaard's writings, always culminate in resignation and solitude.

3. Love is non-rational

To say that love as a passion is partly to identify its non-rational nature. This does not mean that it is irrational to become erotically or romantically involved with another person, only that loving another often defies rational analysis or explanation. There can be no "why" for a love, nothing that offers reasons why one particular person becomes the one you love, and no pattern of reasoned thought that leads to the choice of the beloved. Although a person in love can describe in glowing terms the features of the beloved to whom he or she is attracted, these features are usually more a reflection of the attraction than an explanation or justification of it. That "love is blind" and that lovers, up to a point, will overlook features of the beloved that are offensive or odious only reinforces the observation that love is a passionate attraction defying rational explanation.

In *Stages on Life's Way*, the young man, who confesses that he has never been in love, mentions this odd feature of erotic love in order to ridicule it. He finds it "comic" that, while all human beings want to love and all lovers celebrate their love, no one seems able to explain "what it is that they love" (SLW 34). He continues,

> If erotic love expressed itself in loving the first that comes along, then it would be understandable that one cannot explain oneself more precisely, but since erotic love expresses itself in loving a one and only, a one and only in the whole world, such a prodigious act of separation must in itself contain a dialectic of reasons one would have to decline to hear, not so much because it explained nothing as because it would be too long-winded to listen to. But no, the lover cannot explain anything at all. He has seen hundreds and hundreds of women; he may be getting on in years, has felt nothing. Suddenly he sees her, the one and only—Catherine. Is this not comic; is it not comic that what is going to transfigure and embellish all life—erotic love—is not like a mustard seed that grows to a great tree but, even less, is essentially nothing at all, for not a single antecedent criterion can be stated, as if, for example, there were a specific age in which the phenomenon made its appearance. And not a single reason can be stated as to why he chooses her, her alone in the whole world...." (SLW 36)

4. Love equalizes the lovers

Love demands equality between the lovers. Individuals who are unequal—in social standing, wealth, learning, talents and abilities, or any other attribute—can fall in love. But once they do, they are compelled by their erotic passion to regard one another as equals. One cannot look down on one's beloved, and though one can look up, in the experience of awe and devotion, what is esteemed is the beloved qua beloved, not the possessor of status or other worldly goods. In love, the state of being loved and the ability to love are the essential and defining features of personhood and these exclude all other differentiating criteria.

So powerful is this dynamic of love for Kierkegaard that worldly differences can even become an obstacle to love's expression. In the *Philosophical Fragments* we meet a king in love with a humble maiden who, in order to accomplish his desire to win her, must conceal his royal status and come to her in the form of a commoner. After their marriage, Johannes Climacus observes "… let all be festive while erotic love celebrates its triumph, for erotic love is jubilant when it unites equal and equal and is triumphant when it makes equal in erotic love that which was unequal" (PF 27). Shortly after reading this erotic example, we learn that it has bearing on the love of God for human beings through Jesus Christ. This illustrates how erotic love for Kierkegaard serves as an instrument for conveying larger religious truths. It also suggests how love in all its forms shares common features.

5. Love unites freedom and necessity

Lovers typically believe themselves fated or destined for one another. Their love "has to be" (EO2 20). Understood religiously, this sense of fatedness takes the form of viewing the lover not as one's chosen person but as God's gift. Thus, the Married Man identified as the author of "Some Reflections on Married Life," in *Stages on Life's Way*[7] remarks that the phrase "to choose" in love

---

[7] The Married Man in *Stages* is identified before his lengthy manuscript as Judge William (SLW 82–86). However, so as not to confuse these two, possibly different expressions of pseudonymous points of view, we here refer to the personage of *Stages* as the Married Man.

does not mean "wanting to set someone up as the beloved" but rather "wanting to accept the beloved" and to "thank the god for the gift" (SLW 121). Even the seducer, Kierkegaard's renowned esthete, "lets falling in love stand as something he cannot give himself" (SLW 148). At the same time, this acceptance of the beloved in no way abrogates freedom. Kierkegaard never suggests that God coerces anyone into loving the beloved. Quite the contrary, a man realizes that he loves a woman, chooses her, and then recognizes that God made her known to him. As Judge William puts it, "... in its genius romantic love is free and that precisely ... is what constitute its greatness" (EO2 21). In *Stages on Life's Way,* the Married Man tells us that, "the person who received the gracious gift of immediacy lets himself be married to it in the resolution—and this is indeed the beautiful meaning of marriage" (SLW 148). In *Two Ages* Kierkegaard joins love's themes of passion and freedom when he says: "As is the case with all passion, so it is true of erotic love that the one initiated into it stands free in the consecrated moment of falling in love, free on the rash summit of illusion, free and surveying the whole wide world" (TA 50).

Love thus "unites freedom and necessity" (EO2 45). This "unity of contrasts" (EO2 60) is not unique to erotic love. As Judge William makes clear in *Either/Or,* freely choosing our given life situation is a defining feature of the ethical sphere of existence. Nevertheless, erotic love serves as a special symbol for all of human life because it so intensely expresses our nature as situated yet free beings. Indeed, Kierkegaard's treatments of erotic love suggest that its importance in human life is at least partly owed to the ways in which it permits us to live out and joyfully express these paradoxes of human existence.

6. Love unites time and eternity

Mortal, caught up in time, subject to change and decay, human are also spiritual beings that crave connection with that which endures. Nothing better illustrates the ability of love to express and resolve the paradoxes of human life than the way it brings eternity into time. Poetry and song celebrate the belief that "love is forever." Lovers typically say that they have always loved one another, and that their love will never die. For Kierkegaard, these sentiments are not poetic exaggerations: they convey love's most distinctive and impressive feature. Romantic love, Judge William tells us "… is noble by virtue of the consciousness of the eternal that it assimilates, for it is this that distinguishes all love [*Kjerlighed*] from lust: that it bears the stamp of eternity" (EO2 21). The young man in *Stages on Life's Way* describes the embrace as the confirmation of love because it provides physical expression of the couple's eternal love for one fervent moment in time (SLW 42). In *Either/Or* Judge William deploys this understanding in his defense of married love. The married man, he explains, is by no means lost in the humdrum of household life. On the contrary, he "has not killed time but has rescued and preserved it in eternity. The married man who does this is truly living poetically; he solves the great riddle, to live in eternity and yet to hear the cabinet clock strike in such a way that its striking does not shorten but lengthens his eternity…." (EO2 138). These themes continue to resonate throughout Kierkegaard's authored writings. In *Works of Love*, Kierkegaard asks what it is that "connects the temporal and eternity," and answers "what else but love" (WL 6).

Love's Dialectic

In Kierkegaard's writings, these features of erotic love are integrated within a dynamic and dialectical understanding of love's trajectory in human life. The delineation of this dialectical trajectory constitutes Kierkegaard's most important philosophical contribution to our understanding of love. Movement forward in love's course is required because the features of love, though present in every existence sphere, are less suitably expressed in some existence spheres than others. As a driving force, therefore, love can either propel individuals into despair or force them to leap forward to a new existence-sphere where love's deepest intentionality finds more adequate expression. Passionate yearning leading to self-development through a crisis that precipitates decision characterizes the whole dialectic of existence for Kierkegaard. That erotic love illustrates this dialectic so powerfully once again reveals its importance in Kierkegaard's thinking.

Love in the Esthetic Existence-sphere

The esthetic sphere of existence is marked by the predominance of mood, feeling, and satisfactions or dissatisfactions visited on the individual from without by her or his experiences. Persons whose existence is defined within this sphere find the principle of their being outside themselves. The esthete in love experiences all the features of love we have mentioned, but because of the esthete's essential presuppositions, these features become experiences before which the self is a passive agent: one suffers one's love. Understood in this way, erotic love in the esthetic sphere contradicts itself in many ways. Love can even turn into its opposite. One who is passionately and irrevocably devoted to the lover can also become a helpless victim of the lover's whims. This is exploited by the seducer, who revels in fatally binding another to himself. For the seducer's victim, as is true with Donna Elvira, the timeless devotion of love transmutes itself into an eternal obsession leading to hatred for oneself and one's betrayer (EO1 196).

The esthete savors love's intense joys. The seducer in *Either/Or* confesses that he is dazed by the experience of falling in love and has "gone under in love-rapture" (EO1 324). Yet since the constant flow of satisfactions is his life's task, the esthetically inclined individual simultaneously fears the prospect of attachment to another because it threatens loss of control (EO1 297). One response is cynically to ridicule love. The young man in *Stages* does so when he remarks, "It is comic that erotic love's lofty soaring (wanting to belong to each other for all eternity) always ends up … in the pantry…." (SLW 42). In *Either/Or*, the esthete pokes fun at the widower and widow, each with five children, who "combine forces" and "nevertheless assure one another on the wedding day that this love is their first love" (EO1 254). Yet this derision is really the deepest despair and self-contradiction since no one knows better than the esthete does that erotic love brings life's highest satisfactions. In *Either/Or* Judge William underscores this self-contradiction when he tells the young man, "you do not believe in the eternity of first love…. It is you who so often set yourself up as its knight, and yet you do not believe in it—indeed, you profane it" (EO2 145).

Love in The Ethical Existence-Sphere

This collision within the esthetic existence-sphere—that its highest goal cannot be achieved within its own presuppositions—can either drive lovers to despair or propel them onward to the ethical sphere of existence. The challenge before lovers is to actualize love's deepest promise, especially its ability to offer an intense satisfaction resistant to change. This can be accomplished only through ethical resolve. By accepting the imperative of (marital) commitment, "You shall love," lovers become assured of the continuance of their relationship and the triumph of their love, including its esthetic and erotic dimensions, over all of life's reverses. This in no way excludes and esthetic-erotic dimension of their love. Marriage "presupposes [erotic love] not as something past but as something present" (EO2 36). Beginning in erotic attachment, and then set on a firm basis of mutual commitment, marital love is able to "preserve the esthetic even in everyday life"(EO2 9).

In *Either/Or*, Judge William extensively develops the logic of this position in his letter to the young esthete. The Judge is aware that, superficially regarded, a duty to love seems to be a self-contradiction: a sign that constraint is necessary because spontaneous affection has vanished or will soon do so. The Judge frames the esthete's objection: "You say, 'within itself, marital love is hiding something completely different; it seems so gentle and beautiful and tender, but as soon as the door is shut on the married couple and before one can say Jack Robinson, out comes Master Erik[8]; then the tune is changed to duty'" (EO2 145). The esthete's mistake here is to believe that duty is something external to the couple's love. Instead, duty is their deepest wish, and it is freely accepted by the couple, first in their mutual commitment and then in their marital vows.[9] It also provides the confidence that unending continuance in love can be achieved. Because ought implies can, the lovers' wholehearted acceptance of this duty presupposes their faith that their love can and will abide (EO2 145).[10]

The esthetic fulfillment of erotic love requires ethical resolve, which in itself contains a religious dimension. Lovers perceive

---

[8] A decorated birch branch with which children awaken their parents on Shrove Monday. Here, presumably, the birch of discipline.

[9] Without diminishing the importance of the wedding ceremony, the Judge emphasizes the primacy of each lover's inward commitment to the other, and he affirms that the wedding ceremony "allows what was already in motion to appear in the external world" (EO2 93).

[10] For a more extensive discussion of the logic of the Judge's argument and its possible dependence on Kantian presuppositions, see Ronald M. Green, "Kierkegaard's Great Critique: *Either/Or* as a Kantian Transcendental Deduction" (reprinted here as chapter. 8).

their drive to mutual obligation as a form of divinely given "nourishment" and a guarantee of their love's continuance. Like all ethical resolve, therefore, marital duty rests on religious faith.[11] In *Stages on Life's Way*, the Married Man exalts these implicit religious dimensions of the ethical component of married love. A person, he observes "does not dare cling to himself as a singular individual if he is going to venture out with his love. His comfort is precisely that he is just like other human beings and in this common humanity is in relationship with God by faith and by the resolution" (SLW 164).

Love in the Christian-Religious Existence-sphere

Has erotic love a place in the Christian-religious sphere of existence? Despite the natural assumption that love's dialectical development might continue into this sphere, there are several reasons for believing that erotic love reaches its culmination in marriage and does not carry over from the ethical to the specifically Christian-religious existence-sphere. Paradoxically, the ethically informed religious dimensions of marital love form one of several obstacles to imagining a particular Christian-religious expression of erotic love. For if erotic love can reach complete fulfillment in marital expression, where it rests gently in faith in God's presence as guarantor of the relationship, what need is there for the dramatic crisis and decision that typically mark the transition from one existence-sphere to another? Christian faith, Kierkegaard and his pseudonyms tell us, always involves a leap beyond the ethical sphere of existence precipitated by the experiences of sin, guilt, and repentance (CUP1 267–268; cf. JP1–452; JP4–4011, 4012). But the Judge's vision of marital love

[11] For a discussion of Kierkegaard's understanding of the relationship between ethical resolve and religious faith, see Ronald M. Green, *Kierkegaard and Kant: The Hidden Debt* (Albany: State University of New York Press, 1992) 137–138.

evidences nothing like this. Although repentance has its place in the Judge's concept of the formation of the self (EO2 216–218, 237–238), he does not typically bring sin, guilt, or repentance into relation to the ethical reality of marriage. Marriage's challenge resides for him primarily in the ethical decision to commit to another person. Hence, the religious dimensions of marriage are bound up with the choice of the other that moves erotic love from esthetic immediacy into marital permanence. As the Judge says, "The religious is not so alien to human nature that there must be a break in order to awaken it" (EO2 89).

Of course, neither Judge William nor his alter ego, the Married Man in *Stages on Life's Way* should be taken as representative of Kierkegaard's entire view. Both display, at best, only the earliest expression of the religious (Religiousness A). An alternate view, suggested by *Fear and Trembling*, and seemingly confirmed by some aspects of *Works of Love*, offers a far more discontinuous vision of the fate of erotic love as one moves from the ethical to the Christian-religious sphere of existence. However, this vision also presents obstacles to believing that erotic love has any place in this sphere. Here it is the themes of resignation and self-renunciation that predominate. The lover who chooses God must be prepared to sacrifice any worldly love to the divine command. Abraham evidences this kind of faith when he offers Isaac up as a sacrifice. This is the faith of all other Knights of Faith who, disagreeing with the claim of the Married Man in *Stages on Life's Way* that marriage is "the highest *télos* of individual life" (SLW 101), choose instead to subordinate marital or family obligations to an absolute duty to God (FT 70; CUP1 408–09).

On this understanding, there appears to be no room for erotic love in the Christian-religious sphere of existence. One who lives religiously either retreats to the monastery or chooses, as

Kierkegaard did, to remain unmarried.[12] In this way, the transition to the Christian-religious sphere is decidedly a crisis and a leap, but one that leads wholly away from erotic love in any of its forms.

Works of Love can be read to support this view. For example, its criticism of the exclusivity of erotic love and its insistence that Christian love is inclusive, taking every neighbor for its object, appears to erase one of the essential presuppositions of erotic love. "Christian love," says Kierkegaard, "teaches us to love all people, unconditionally all. Just as unconditionally and powerfully as erotic love intensifies in the direction that there is but one and only one beloved, just as unconditionally and powerfully does Christian love intensify in the opposite direction" (WL 49). The insistence that true love is selfless and the repeated assertion that beneath a veneer of passionate selflessness, all erotic and marital love is merely secret, enhanced, and augmented self-love, also seems to eliminate any foothold for particular relationships. "*Erotic love and friendship are preferential love and the passion of preferential love*; Christian love is self-denial's love...." (WL 52). In view of these rejections of the essential presuppositions of erotic love, it would seem that anyone committed to the kind of Christian love extolled in *Works of Love* must be prepared to avoid marriage, or, if married, must await its eschatological dissolution. "In the resurrection, they neither marry nor are given in marriage" (Matt 22:30; Luke 20:35).

---

[12] Kierkegaard writings contain statements of respect for monasticism's religious intensity—its "passion and respect for the absolute *télos*" (CUP1 414)–along with criticisms of its effort to leave the world and establish outward signs of what must properly be expressed as religious inwardness within the world: "The middle Ages wanted a little cubbyhole in order to be able to occupy itself properly with the absolute; but it was precisely by this that the absolute lost, because it still became something outward" (CUP1 408).

An Alternate View

As compelling as this negative picture of erotic love's continuance in the Christian-religious sphere might be, however, there are statements throughout *Works of Love* that call it into question and hold out a possibility of Christian love *united with* ongoing romantic-erotic attachment to another person. To use an image borrowed from *Either/Or*, this view sees religious love not as eccentric to erotic love, but concentric with it (EO2 30). For example, after urging the reader to "take away the distinction of preferential love so that you can love the neighbor," Kierkegaard continues:

> But you are not to cease loving the beloved because of this—far from it. If in order to love the neighbor you would have to begin by giving up loving those for whom you have preference, the word "neighbor" would be the greatest deception ever conceived. Moreover, it would even be a contradiction, since inasmuch as the neighbor is all people surely no one can be excluded—should we now say, least of all the beloved? (WL 61)

Again:

> No, love the beloved faithfully and tenderly, but let love for the neighbor be the sanctifying element in your union's covenant with God. (WL 62)

And again:

> Christianity ... knows only one kind of love, the spirit's love, but this can lie at the base of and be present in every other expression of love. (WL 146)

In keeping with this conception, religious love does not put an end to worldly attachments, including intense erotic-romantic love for another person or marital devotion. Christians are no more required to renounce a loving, committed relationship with another person than they are to give up food or drink (WL 47, 52).

Indeed, "the doctrine about love for the neighbor" is specifically intended "for transforming erotic love and friendship" and is meant to "permeate everything" (WL 112). "Christianity does not want to make changes in externals; neither does it want to abolish drives and inclination—it wants only to make infinity's change in the inner being" (WL 139).

Features of Religious Erotic Love

What, then, is this change? What form does erotic love takes in the religious existence-sphere? What changes does it undergo that distinguish it from the ethical expression of love in marriage, as depicted by Judge William or the Married Man? In seeking the answers to these questions we can be guided by an understanding of the basic dialectic that governs all movement from existence-sphere to existence-sphere in Kierkegaard's thought. Thus, the transition from the ethical to the religious stage of erotic love will be abrupt: a leap rather than an evolution. It will be precipitated by a collision or crisis: something within its earlier presuppositions that threatens to drive it to the edge of despair or that undermines its highest aspirations. It will pass through choice and resolve and emerge to rest on essentially different premises than the forms of love that preceded it. Yet, it will be in continuity with these previous forms and a completion of them. Just as the ethical stage of love, marriage, seeks always to retain the esthetic-erotic dimension of love and even intensifies it by adding duration and continuity to the erotic's passion, so might erotic love in the religious existence-sphere continue and intensify love's erotic and ethical elements. It will be, as Sylvia Walsh, drawing on Judge William's suggestive phrasings says, a passionate but deeply reflective "second immediacy."[13]

---

[13] Walsh, "Forming the Heart: The Role of Love in Kierkegaard's Thought" 244. George Connell critiques the Judge's position here, arguing that his vision of marriage as protecting and sustaining immediate erotic love "makes

Against this background, we can directly approach the question of erotic love in the Christian-religious existence-sphere by asking what could possibly upset the kind of resolved marital love sketched by defenders like Judge William and the Married Man? A variety of answers present themselves and all involve causes that erode or challenge the basis of marriage's mutual commitment. Most obvious among these is infidelity on the part of either partner, a breach that assaults the other's pride and sense of justice and that inevitably saps his or her resolve to remain in relationship. Related to this, and equally painful, is the sense that the beloved has become hostile or alienated. Another threat is changes in the beloved (or the self) that are extensive enough to raise the question of whether either is any longer really the same person. Although the importance of certain qualities in the beloved at the start of an erotic relationship may be more a reflection of the love than an explanation for it, fundamental changes in the beloved's most attractive features (or their conversion into their opposite) can shatter resolve by raising the question of whether the lover really is the same person to whom one has pledged eternal devotion. Similarly, deep changes in one can undermine the foundations of one's love for the other. Finally, either partner's freedom can threaten the resolve to faithfulness. Although mutual freedom, as we have seen, is the premise of all erotic love, freedom and independence also are a deep threat to love because, in freedom, the lover or the beloved can change nature, rebel against, and even reject the love relationship.

How, then, can love survive such assaults? How can erotic love's deep inner intentionality toward continuity be preserved before this host of troubling transformations in oneself and the object of one's love? In *Either/Or* Judge William seems to be unaware of these problems. He and his wife live in untroubled

---

demands altogether exceeding human capacity." See his *To Be One Thing* (Macon, Georgia: Mercer University Press, 1985) 179.

unity. They hold no secrets from one another, even to the extent of confessing to one another breaches of fidelity (EO2 117, 119, 179).[14] Against this too simple vision, *Works of Love* can be read as dealing with the real stresses that threaten marital continuity and offering a response to them with a vision of religious love that penetrates and transforms, but does not necessarily abolish, erotic attachment. We can see this by looking closely at some of the distinguishing features of the neighbor-regarding love Kierkegaard describes there in order to see how it addresses the assaults that threaten even the most devoted and passionate love relationship.

It Is Not Based on Preference

One distinctive feature of the religious love that Kierkegaard depicts is that it is not focused on what distinguishes the beloved, on what Kierkegaard calls "passionate preference," but on the beloved's essentially human qualities. Kierkegaard terms this "the eternal equality in loving" (WL 58). "In being king, beggar, rich man, poor man, male, female, etc., we are not like each other—therein we are indeed different. But in being the neighbor we are all unconditionally like each other" (WL 89). This essential perception of likeness rather than difference, Kierkegaard insists, applies also to the marital relationship:

> Your wife must first and foremost be to you the neighbor; that she is your wife is then a more precise specification of your particular relationship to each other. But what is the eternal

---

[14] In "Remaining True to the Ethical? A New Letter from Assessor Vilhelm, with Commentary," (Paper delivered at Søren Kierkegaard Research Centre Conference, Copenhagen, 17 August 2007), George Pattison imagines a broken and distraught Judge William who authors a further paper to the young man. This Vilhelm's marriage has fallen apart as a result of his overbearing, masculinist efforts to control his wife's freedom and self-development.

> foundation must also be the foundation of every expression of the particular. (WL 141)

Superficially regarded, this placement of the beloved on a plane of equality with others would seem to be a demotion in status. For how could being seen to be *like others* be superior to being the extolled object of erotic love and preference? Yet, when we consider that one of the greatest threats to erotic love is precisely the perception by one party to a relationship that the other has changed or lost those features that cemented preference in the first place, we see the importance for love's continuance of what Kierkegaard is saying here. *"The task,"* Kierkegaard reminds us, *"is to find the once given and chosen object—lovable, and to continue to find him lovable no matter how he has changed"* (WL 159). We do this by loving the (actual) person we see, by accepting her or him as they are and not as we would have them be, and by recognizing that they are an imperfect creature like ourselves and all other human beings. In these ways love is strengthened against all the deceptions and disappointments that preference can undergo. In this way, as well, love achieves the unconditional acceptance of the other to which marriage aspires but which its basis in preference sometimes threatens. Many statements in *Works of Love* reinforce this point:

> To be able to love a person despite his weaknesses and defects and imperfections is still not perfect love, but rather this, to be able to find him lovable despite and with his weaknesses and defects and imperfections. (WL 157–58)

> When it is a duty in loving to love the people we see, then in loving the actual individual person it is important that one does not substitute an imaginary idea of how we think or could wish that this person should be. (WL 164)

> The Christian point of view, however, is that to love is to love precisely the person one sees. The emphasis is not on loving the perfections one sees in a person, but the emphasis is on loving the person one sees, whether one sees perfections or imperfections in the person, yes, however distressingly this person has changed, inasmuch as he has not ceased to be the same person. He who loves the perfections he sees in a person does not see the person and therefore ceases to love if the perfections cease, when the change begins, although this change, even the most distressing, still does not mean that the person ceases to exist. (WL 173)

Of course, all erotic love purports to love the *real person* who is the beloved. Marital love trumpets its acceptance of the beloved "for better or for worse" and it cherishes its knowledge of the other's most intimate and embarrassing secrets and vulnerabilities. Yet, what Kierkegaard is suggesting here is that erotic love's greatest fear is precisely the simple humanity of the beloved: his or her likeness to others and real possession of human (as opposed to enchanting) imperfections. By placing this shared humanity to the fore, and by resolving to love the beloved for this shared humanness more than for all the features that elicit passion, Christian love therefore fulfills what erotic love promises. It does so however, in a way that must initially shock the lover's sensibilities and that is arrived at, not through a natural evolution of love, but through the resolve to continue loving despite the stresses to which preference is subject.

It may be objected that Christian love, by seeing the common humanity—the neighbor—in the beloved removes one of the essential features of erotic love: its exclusiveness. Yet Kierkegaard is quick to state that even when seen as a human being, as one neighbor among others, the beloved remains fully the beloved, fully the special and chosen one. Religiously informed erotic love is at once intensely universal *and* intensely particular. Indeed, to distinguish these two aspects of the beloved, to see the beloved first as the chosen one and then as just another human being, or in some respects as the beloved and in other respects as the neighbor, is to artificially separate elements that must completely commingle in a single, multi-dimensional perception:

> The beloved, the friend, is of course a human being also in the more ordinary sense and exists as such for the rest of us, but for you he should exist essentially only as the beloved if you are to fulfill the duty of loving the person you see. If there is a duality in your relationship so that to you he is just partly this individual human being in the more ordinary sense, partly the beloved in particular, then you do not love the person you see. Instead it is as if you had two ears in the sense that you do not, as is normal, hear one thing with both ears but hear one thing with one and something else with the other. (WL 165)

Kierkegaard continues by observing that any such separation in one's perception of the beloved amounts to a critical self-withholding that is incompatible with love. Somehow, we are to avoid this dichotomy of consciousness: While recognizing that the beloved is a human being like all the others we are called to love, we must also never forget that this is our special one. Particularity and universality, exclusivity and openness, are not opposites but somehow mutually reinforcing. Just as we should love the beloved with the honest and penetrating regard that accompanies our love for every human being, so we should blindly love "every human being as the lover loves the beloved" (WL 69).

An illustration from popular culture may help make this difficult point clearer. In the film *Gone with the Wind,* there is an episode where Melanie Wilkes and Scarlett O'Hara are shown serving long hours as volunteers in a Confederate military hospital. Near the end of the day, Melanie makes one more effort to comfort a wounded soldier. When Scarlett asks Melanie how she can summon up the strength to do this, Melanie replies, "This might be Ashley." This remark, by the film's most obviously saintly figure, illustrates the intimate union of universal and erotic love that Kierkegaard suggests. For Melanie, the wounded soldier/neighbor *is* the beloved: his countenance elicits from her the passionate regard she feels for Ashley Wilkes. But from this moment forward, Ashley himself will also evoke for Melanie all the wounded soldiers she has comforted. Erotic love's exclusivity has been simultaneously intensified and transcended.

Finally, we might add that to *see* the beloved as he or she is, with all the defects of a human being (including, presumably the propensity toward sin), does not mean that we must *accept* those defects without struggling to change them. As M. Jamie Ferreira points out, "Love is not ... only accepting; it is also *challenging*."[15] Kierkegaard tells us that "The relationship itself will with integrated power fight against the imperfection, overcome the defect.... The two are to hold together all the more firmly and inwardly in order to remove the weakness" (WL 166–67). Thus, erotic love in the religious stage evidences another paradox: unconditional acceptance of the beloved conjoined with a moral will toward the beloved's betterment. In this sense, erotic love provides a profound illustration of a repeated theme in Kierkegaard's writings: the importance of grace *and* moral striving (JP6–6801; JP2–1878).

It Is Truly Selfless

For erotic love to be a suitable instrument of Christian love, it must first be purged of the egoism that threatens all intense, dyadic love relationships. Lovers typically celebrate their selfless devotion to one another, and no one can doubt that this self-transcendence is one of love's greatest joys. Yet, with keen psychological insight, Kierkegaard observes that the intense other-relatedness and seeming selflessness of erotic love frequently cloaks an intense *égoïsme à deux*:

---

[15] Ferreira, "Equality, Impartiality, Moral Blindness in Kierkegaard's *Works of Love*" 82.

> [E]rotic love and friendship are the very peak of self-esteem, the intoxicated *I* in the *other I*. The more securely one *I* and another *I* join to become one *I*, the more this united *I* selfishly cuts itself off from everyone else. At the peak of erotic love and friendship, the two actually do become one self, one *I*. This is explainable only because in preferential love there is a natural determinant (drive, inclination) and self-love, which selfishly can unite the two in a new selfish self. (WL 56)

Here too, Christian love proves liberating for those in a relationship. However intoxicating this conjoint egoism may be, it ends by isolating the pair from others outside the relationship. With each partner's ego at stake, any hint of defection by one partner becomes a personal threat to the other. The most poisonous fruit of this mentality is jealousy. "Jealousy loves as it is loved ... Anxious and tortured by preoccupation with itself, it dares neither to believe the beloved absolutely nor to give itself wholeheartedly" (WL 35). At this extreme, love's bond becomes a prison for each member of the pair. Against this, Christian love, by insisting on love for *all* one's neighbors and by demanding a genuine selflessness, breaks the chains of this intoxicating dyadic egoism and frees each party for self and others:

> The spirit's love, in contrast, takes away from myself all natural determinants and all self-love. Therefore love for the neighbor cannot make me one with the neighbor in a united self. Love for the neighbor is love between two beings eternally and independently determined as spirit; love for the neighbor is spirit's love, but two spirits are never able to become one in a selfish sense. (WL 56)

God Is Present as a Third

Christian love also transforms and sustains erotic love by admitting God, as a third person, into the lover's relationship:

> *Worldly wisdom is of the opinion that love is a relationship between persons; Christianity teaches that love is a relationship between: a person—God—a person, that is, that God is the middle term.* However beautiful a relationship of love has been between two people or among many, however complete all their desire or all their bliss have been for themselves in mutual sacrifice and devotion, even though everyone has praised this relationship—if God and the relationship with God have been omitted, then this, in the Christian sense, has not been love but a mutually enchanting defraudation of love. *To love God is to love oneself truly; to help another person to love God is to love another person; to be helped by another person to love God is to be loved.* (WL 106–107

Once again, Kierkegaard's vision of Christian love seems to assault the very foundation of erotic love. Passionate love for another, we have seen, is by its nature exclusive. How can it admit a third party, even God, without distracting the lovers' passionate attention from one another? How can it admit a goal for each party to the relationship—love of God—that seems to draw attention away from the beloved? That Kierkegaard himself appears to have sacrificed his relationship with Regine in order to fulfill his religious vocation only points up the problem.

If we keep in mind the nature of erotic love and the perils that confront it, however, we can see in what senses this insistence on the presence of God in an erotic relationship is the very condition for its vitality and continuity. First of all, the presence of God is the antidote to worshipful idolatry of the beloved, "[T]he relationship among human beings ought and may never be such that one worships and the other is the one worshipped" (WL 125). Although such worship is erotic love's very instinct, it must lead inevitably to disappointment and despair, for no human being is the god that love's enchantment initially perceives. Christian love avoids this by fixing attention on God and on each member of the pair's relationship with God.

God's presence also insures the independence and freedom of each partner. Erotic love, we recall, is predicated on free choice of it by each of the parties. Although "made for one another," the lovers nevertheless choose one another freely. And yet, the perceived fatedness of love also qualifies freedom in many ways, not least of all each lover's unwillingness to see the beloved exercise freedom in ways that imperil the relationship. "You are free to do anything but reject me, reject us," is every lover's plaintive cry. By reminding us that each person stands in an independent relationship to God, Christian love creates the space, the breathing room—the "air"—as Kierkegaard might say, to which erotic love aspires but which it simultaneously resists:

> [I]s it actually love, in the divine sense, to show a devotion such as the object of love demanded? Next, is it love, in the divine sense, on the part of the object of love to demand such devotion? Every person is God's bond servant; therefore he dare not belong to anyone in love unless in the same love he belongs to God and dare not possess anyone in love unless the other and he himself belong to God in this love.... If there was between two or among several a relationship of love so happy and perfect that the poet was bound to exult in it ... this is by no means the end of the matter. Now Christianity steps forward and asks about the relationship to God, whether each individual is first related to God and then whether the relationship of love is related to God. If this is not the case, then Christianity, which certainly is the protector of love or because it is that, in God's name will not hesitate to split up this relationship until the lovers are willing to understand this. And if only one party understands it, then Christianity, which certainly is the protector of love, will not hesitate to lead him out into the horror of a collision such as no poet dreams of or has ventured to portray...: out of love and in love *to hate the beloved*.... Christianity does this not merely to collect, as it were, God's outstanding claim ... but does it also out of love for the lovers, because to love God is to love oneself, to love another person as

> God is to deceive oneself, and to allow another person to love one as God is to deceive this other person. (WL 107–108)

As this passage suggests, love for the other can even require doing that which appears to harm the beloved and be motivated by hate. This can occur when the beloved must be rudely awakened from the bliss of devotion to the reality of his or her own independence, freedom, and responsibility. Since any attempt to explain or justify this "tough love" (to use a modern phrase) may only cause the beloved to love you even more (WL 114), a sincere effort to awaken the beloved may lead unavoidably to a breach of the relationship. When there are only two parties in a relationship and when each is the other's supreme goal, truly loving the other in this way becomes impossible because it violates the relationship's essential condition, which is to please the beloved. "[S]uppose the beloved saw that the relationship would become the lover's ruin, would completely shake his distinctiveness—well, then erotic love as such does not have the power to make this sacrifice" (WL 273). But when God is present as the third person in a love relationship and when, as Kierkegaard says, the two are related through God to the reality of loving and not merely to pleasing one another, then whatever truly is needed for the beloved's good can be accomplished.

Not only does God's presence make this difficult but loving resolve necessary (because each partner is independently related to God), but it simultaneously sustains the confidence that whatever the beloved may think or do, and whatever the world may conclude, one's conduct is nevertheless truly an expression of love. It is the tragedy of such efforts aimed at helping another person to exercise his or her freedom that these efforts must to some extent be "invisible." The goal is to help the other person become his or her own master. But, as Kierkegaard points out, to reveal one's involvement in accomplishing this contradicts the aim. For if the person helped sees that they have actually become his or her own master through another's help, then "the person helped has not become his own master" (WL 279). In a certain sense, therefore, when love succeeds in freeing its object, the one who accomplishes this will outwardly appear to have completely squandered her or his life on the existence of another. But the genuine lover draws strength from the presence of God in their relationship and rests content in "simply being an active power in the hands of God" (WL 279).

Finally, God's presence as the third party in a love relationship provides confidence that the love will survive any breach even if the beloved chooses to terminate contact. This is so because the lover is oriented not merely to the consciousness of the beloved but toward God and to love itself:

> When love ceases, when ... in the loving relationship between two people something comes between them so that love ceases, then the two, as we human beings say, break up.... Christianity, however, does not know this use of language, does not understand it, refuses to understand it.... When a relationship is only between two, each one always has the upper hand in the relationship by being able to break it, because as soon as one of them has made the break, the *relationship* is broken. But when there are three, no one of them can do it. The third ... is *love* itself, to which the innocent sufferer in the break can then hold—then the break has no power over him. (WL 303–305)

God's presence in a love relationship is thus the guarantee of continuity. But continuity—eternity—we have seen, is love's deepest ambition, from its erotic expression forward. Judge William and his alter ego, the Married Man, who reflect the ethical sphere of existence, appear to believe that continuity can be attained through ethical resolve. In *Works of Love,* Kierkegaard does not deny the importance of such resolve. "You shall love," remains the banner under which love proceeds in the religious existence-sphere (WL 17–43). But what Kierkegaard adds to this ethical resolve is the understanding that unless God, and through God, the beloved *as neighbor*, become central to the lovers' relationship, even the most committed erotic love will be pulled asunder.

Conclusion

We have tried to show that Christian love can play a decisive role in sustaining committed erotic love in the face of the inevitable assaults that menace it. It does so in three ways: 1) by drawing attention to what is essentially human in the beloved and thus freeing each lover from preoccupation with the beloved's ever-changing particularity; 2) by chastening the augmented egoism that can makes a love relationship destructive both within and without and that fuels jealousy and resentment; and 3) by placing each party in independent relationship to God so as to provide room for personal development, freedom, and courageous mutual support.

Within the context of a loving erotic relationship or marriage none of these insights is easily achieved. Love's passionate attachment to the other (and fearful clinging to the relationship for one's own sake) resist opening the duo to a third, even God. The freedom of the beloved, though a presupposition of all love, is also love's gravest threat and is typically resisted in many ways. The marital vows can themselves be used to imprison and bind the other to a destructive relationship. Thus, ethical resolve and commitment is not enough. The promise of erotic love is fulfilled only when each party's independent humanity is acknowledged, including each party's propensity toward sin and guilt. This last point is important: sin in oneself; sin in the beloved are erotic love's supreme foes. Sin as infidelity and betrayal by the other are humanly insupportable and fracture all erotic relationships or turn them into hatred. Sin in oneself, as the tendency to exalt the lover in order to control him or her—professing selflessness in the name of extreme selfishness—is the subtle undoing of other-regard. These are the final obstacles that erotic love must overcome to fulfill its inner drive toward eternity and genuine love for the other. But for this to happen, Kierkegaard tells us, God must enter the relationship.

This suggests that erotic love in the religious existence-sphere, like all such decisive transitions, is achieved only by a leap. It is not part of a seamless evolution from marital commitment, but is born in the crisis of some of the ethical view of marriage's more naive assumptions. Judge William in *Either/Or* and the Married Man of *Stages on Life's Way*, in their untroubled marital bliss, are still on their honeymoon. When they confront in themselves or their beloved aging, illness, ennui, betrayal, or most threatening of all, the development of a free and independent self, their feelings about love or their beloved may change. The ethical framework in which they live, premised on both parties' untroubled fidelity to their idealistic marital vows, has little room for the genuinely unconditional acceptance of the other that is the essence of erotic love in all its forms.[16] Anticipating this, Kierkegaard offers in *Works of Love* a guide to erotic love's survival. Having passed through esthetic and ethical crises, erotic love in its religious expression finally achieves its promised passion for the beloved, its true equality, its freedom for each party in the love relationship, and its eternal unchangeability. In this sense, *Works of Love* is not just a study of Christian love. It is the capstone of Kierkegaard's entire philosophy of love.

---

[16] "In erotic love and friendships, we rightly insist that the other's affection for us may not in any way be reducible to any moral algorithm, even if our moral behavior is a factor. And our insistence reflects a human value that is independent of intrinsic moral value."—Laurence M. Thomas, "The Fragility of the Moral Self: Self-Love and Morality," Poynter Center Monograph Series, *Essays on Human Institutions* (Bloomington, Indiana: The Poynter Center, October 1997).

Index